MASTER EXAM WORKB

By Tom Henry

Based on the 2005 National Electrical Code®

This workbook contains fourteen closed book exams and sixteen open book exams for a total of over **1820** electrical exam questions with answers and references. This workbook is designed to help prepare the electrician for the Master electrical examination.

Fourth Printing October 2006

ISBN 0-945495-75-7

HENRY PUBLICATIONS SINCE 1985

Introduction note from Tom Henry.......

I have been involved with electrical examinations for over 40 years and I would like to share the following with you.

I've heard all the negatives from the applicants over the years and I can't say that I don't agree. Comments such as:

- •Why don't they test me on what I do?
- •Why can't I take reference books with me?
- •Why don't they test me on my skills rather than my book knowledge?
- •Why can't I see my exam and learn from the questions I missed?
- •Why are exams written by engineers?
- •Why are politics involved?
- •Etc, etc, etc.

Examinations today are based on your cognitive ability rather than your hands on skill level. Mainly because the exams are computer graded due to the volume and time required. Exams are written at different levels from fairly easy to very difficult depending on the requestor.

As an electrician today you are in a very specialized field. You may work in residential whereas another exam applicant is working in the fields of Kansas on oil well equipment. The exam questions come from all areas of the electrical industry that's why so many applicants must learn the entire picture in order to be successful on an electrical examination.

I find so many applicants are looking for a short cut, there is NO short cut, as there is no ONE book that will prepare you. You have chosen a skilled trade that without proper training and education has caused injury, property damage and death. I have found that being an electrician requires a continuing education throughout your entire career.

Today, building departments are switching from one testing company to another frequently. Electricians today are on the move, next year you may be working in a different city, county, state or even a different country. In most cases you will be required to take another examination. Today an electrician must keep up with electrical education as new products and technology are changing everyday.

My final comment is prepare for the exam properly. There are only three areas, closed book, open book and calculation questions. Prepare for all three properly. Today some exams state "open book" questions. but if you note they ask questions that are NOT from the Code book but on theory, general knowledege, AC-DC fundamentals, controls, etc. BUT, the only reference book allowed in the exam room is the Code book.

You MUST prepare for the "NO BOOK" questions which I refer to as CLOSED BOOK. This is a mistake I see many applicants making today. Also, remember that where you are scheduled to take your exam may switch formats. So always prepare fully with closed book, open book, and calculations.

There is no penalty for knowing too much!

This book was written as a study guide to prepare for a Master Electrical Examination. Different states, cities and counties have different exam formats and qualifying rules and requirements.

Inquire with your local building department with questions pertaining to an electrical license examination. Ask for a registration and scheduling instructions. As you need to know the examination outline. Example: What date is the exam given, how many questions, multiple choice, subject area, time of the exam, what reference books are allowed in the testing area, etc.

Most electrical exams are multiple choice and the subject areas of testing is basically the same. The applicant is expected to be able to answer general knowledge questions without a reference book. These type of questions are referred to as "closed book questions" or "no book questions" regardless of how they are titled the applicant is required to know the answer without a book.

A new type of electrical exam has surfaced in which the Master is asked 100 questions with a four-hour time limit. The heading of the Examination Outline is **OPEN BOOK**. The only reference permitted in the testing area in the National Electrical Code of which appromimately 74% of the open book questions are asked. Where I get confused is on an **OPEN BOOK** exam approximately 26% of the questions are asked **without** the use of a reference book. General knowledge questions from the American Electrician's Handbook, Alternating Current Fundamentals and Industrial Motor Control books which are **NOT** allowed in the testing area. Questions from theory, safety, transformers, etc.

These "no book" questions formerly were titled "closed book" questions, either way, you are required to know the answer without the use of a book.

OPEN BOOK QUESTIONS 74%

NATIONAL ELECTRICAL CODE ONLY BOOK ALLOWED

GENERAL KNOWLEDGE QUESTIONS 26%

NO REFERENCE BOOKS ALLOWED

Remember, you must become familiar with the type of exam you are going to take. Make sure you know what subjects are covered on the exam. Find out all you can, how many questions will be asked? What type of questions, multiple-choice, true-false, fill in the blank? Show your work or computer graded? What is the time allowed for the exam? What score is required to pass? These are some of the questions that will help you in your preparation. Get a blueprint of the exam you are to take.

This book **"Master Electrican"** with exam questions and answers will prepare you for both, the code book related questions and the new "no book" (formerly titled closed book) questions. You will also need to study others books on calculations, Ohms Law, theory, transformers, motor controls, etc. in your proper preparation for the electrical examination.

•This exam format is now being adopted by several building departments throughout the U.S.

I.C.C. EXAMINATIONS OUTLINES

(I.C.C. is the International Code Council)

Master Electrician and Journeyman Electrician outlines and examinations are the same for all jurisdictions in the Texas Standard Program.

JOURNEYMAN ELECTRICIAN

ONE PART OPEN BOOK 80 QUESTIONS 4 HOUR TIME LIMIT

MAJOR CONTENT AREA	NO. QUESTIONS	% OF EXAM
Services & Service Equipment	13	16.25%
Wiring Methods & Installation	19	23.75%
Cabinets, Panelboards, Switchboards, Boxes, Conduits	4	5%
Conductors	16	20%
Control Devices	3	3.75%
Motors & Generators	6	7.5%
Utilization Equipment & Devices	6	7.5%
Special Occupancies & Use	9	11.25%
Miscellaneous	4	5%

MASTER ELECTRICIAN

ONE PART OPEN BOOK 100 QUESTIONS 5 HOUR TIME LIMIT

MAJOR CONTENT AREA	NO. QUESTIONS	% OF EXAM
Services & Service Equipment	19	19%
Wiring Methods & Installation	16	16%
Cabinets, Panelboards, Switchboards, Boxes, Conduits	4	4%
Conductors	14	14%
Control Devices	5	5%
Motors & Generators	10	10%
Utilization Equipment & Devices	7	7%
Special Occupancies & Use	10	10%
Miscellaneous	5	5%
Plan Reading & Analysis	10	10%

Master Electrical Exam

•This is a format now being used in some areas of the United States.

License Information

Scope - Tests a candidate's knowledge of the design, plan, layout, installation, repair and alteration of electrical conductors, fixtures, appliances, apparatus, raceways, conduit and related equipment and fixtures that use electrical energy for light, heat, power, data and communications.

Examination Outline

Open Book
100 Questions - Four-hour time limit
75% Correct Required to Pass

SUBJECT AREA	PERCENTAGE
GENERAL ELECTRICAL KNOWLEDGE & CALCULATIONS	10
RACEWAYS & ENCLOSURES	10
SERVICES, FEEDERS & BRANCH CIRCUITS	10
OVERCURRENT PROTECTION	5
CONDUCTORS & CABLES	9
GROUNDING & BONDING	16
UTILIZATION & GENERAL USE EQUIPMENT	9
SPECIAL OCCUPANCIES	5
SPECIAL EQUIPMENT & CONDITIONS	5
MOTORS & CONTROLS	12
LOW VOLTAGE & COMMUNICATION CIRCUITS	6
SAFETY	3

References

The reference materials listed below should be helpful to candidates in preparing for this examination. These materials may **NOT** contain all of the information needed to be competent in this trade to pass the examination. We try to keep this reference list current and consisting of industry-standard materials, However, due to circumstances beyond our control, some of these references may go out of print or be unavailable. Please contact a bookstore 1-800-642-2633 for similar reference materials. This reference list is only a small sample of materials available which contain the necessary information.

The following reference IS allowed in the testing center:
NFPA 70-National Electrical Code.

The following references have been used to create exam questions but are NOT allowed in the testing center:
American Electricians' Handbook, Alternating Current Fundamentals, and Industrial Motor Control.

MASTER ELECTRICAL EXAMINATIONS
GENERAL INFORMATION

•This format is very popular. It has been used for over 40 years. It's best to prepare for this format as it covers all areas of an electrical examination.

Each examination consists of three parts. Parts I and II are made up of multiple-choice questions. These two parts count fifty percent of the total examination grade.

PART I (Closed Book - One Hour) includes definitions, general information, code questions of the type the electricians would use in their daily work and basic theory. This part contains 50 questions.

PART II (Open Book - Two Hours) includes specific code questions taken from the National Electrical Code and additional theory questions. This part of the Master Electrician examination contains 70 questions.

PART III (Open Book - Three Hours) consists of 30 multiple-choice problems. This counts 50 percent of the final grade. Applicants will be required to solve problems from the following categories, using the data provided in the questions.

1. Residential load
2. Conduit fill
3. Demand load
4. Voltage drop single phase
5. Motor, branch circuit, feeder sizes and protection single phase
6. Ampacity and load current
7. Transformers
8. Appliance loads
9. Service load problem (commercial)
10. Voltage drop three phase
11. Three phase 240/120v high-leg delta loads
12. Motor, branch circuit, feeder size and protection three phase
13. Motor control
14. Others

If there is a difference between the National Electrical Code and a local code, the applicant is expected to use the National Electrical Code on Parts I, II and III.

PREPARING FOR A CLOSED BOOK EXAM

Part I Closed Book

This part of the test is where common sense, apprenticeship and years on the job are helpful. Safety type questions are asked, questions on practical knowledge as the proper connections to a switch circuit. Ohms law and basic theory questions are asked in Part I closed book. If they want to make the closed book exam more difficult, they ask questions from the Code book. Definitions from Article 100 are a favorite closed book question as they expect the electrician to know the definitions. Prior to the exam the last thing the applicant should do is scan Article 100 in the Code. Try to retain as many definitions as you can. The first part of the exam that you are handed is Part I which contains definitions closed book.

A time limit of one hour is allowed to answer the 50 closed book questions. The 50 questions can be answered easily in less than an hour. It's very simple, either you know the answer or you don't. It doesn't help to sit and scratch your head pondering over the correct answer. It has been proven in test taking that the longer you hesitate in selecting the choice, the more likely you are to talk yourself out of the correct answer.

Read the question and the choice of answers **carefully** and select your choice and move to the next question.

After applying the work required by this book you will be able to answer the 50 questions in twenty to thirty minutes. When Part I is completed raise your hand and ask for Part II of the exam. This will provide you with extra time for Part II open book which, I feel, is the most difficult part of the exam.

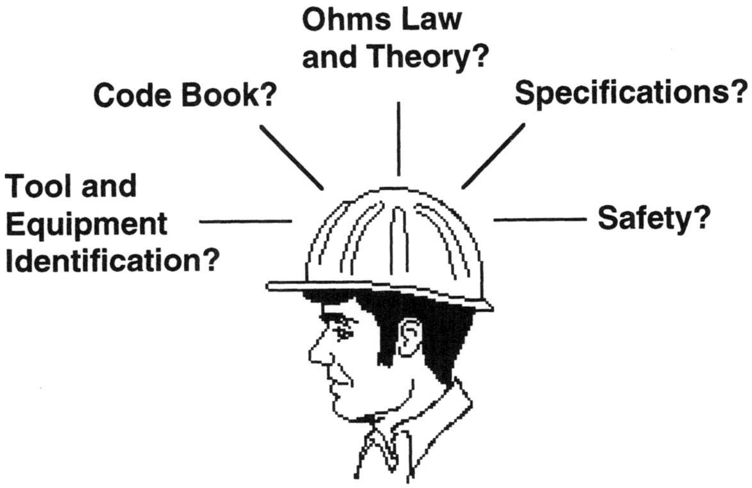

How much do you know, or can remember of the subjects asked?

Excellent study aids to prepare for the closed book exam are the audio tapes Item #192 and the playing cards Item #392.

MASTER CLOSED BOOK EXAM 50 QUESTIONS

Shown below is how the 50 questions are divided up.

- 30-32 questions are asked from general and applied theory which represents 62% of the exam
- 7-9 questions are asked from Chapter 1 which represents 16% of the exam
- 4-6 questions are asked from Chapter 2 which represents 10% of the exam
- 0-2 questions are asked from Chapter 3 which represents 2% of the exam
- 1-3 questions are asked from Chapter 4 which represents 4% of the exam
- 1-3 questions are asked from Chapter 5 which represents 4% of the exam
- 0-2 questions are asked from Article 90 which represents 2% of the exam

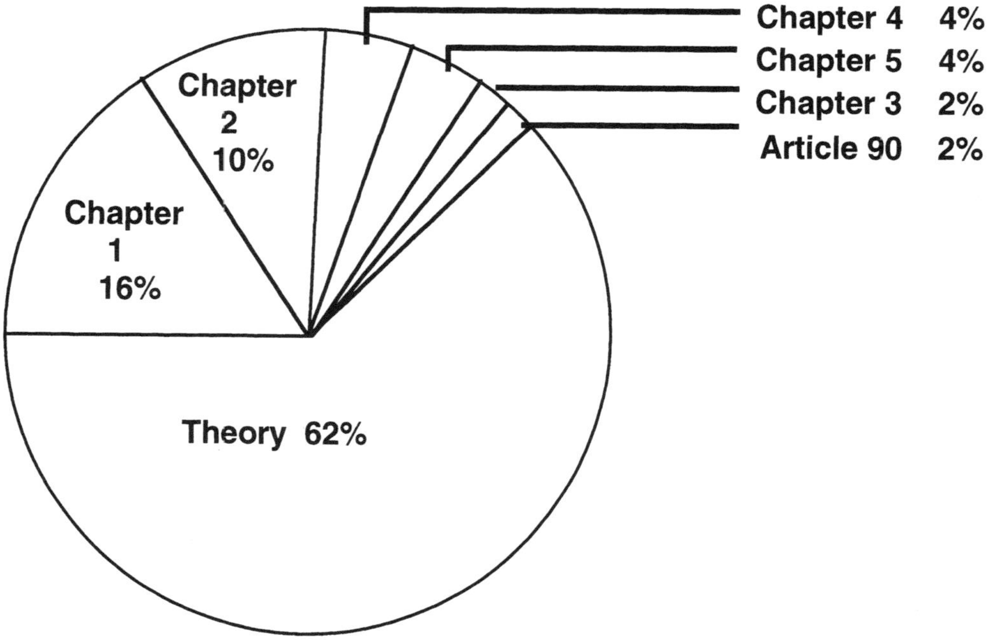

Theory is a big part of an electrical exam. The Ohms Law book and Theory book by Tom Henry are a must!

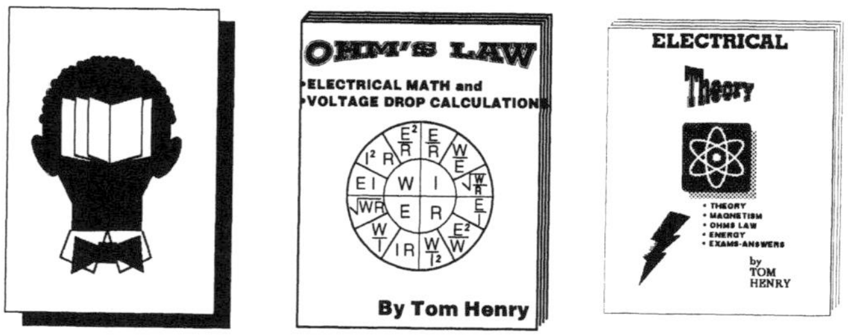

CLOSED BOOK (or no book) EXAM QUESTIONS - Practice

1. When testing a circuit for continuity using the multimeter, the electrician should ____.

(a) insert the test leads and connect the test leads in parallel with the circuit after energizing the circuit
(b) insert the test leads and connect the test leads in series with the circuit after energizing the circuit
(c) insert the test leads and connect the test leads in series with the circuit after de-energizing the circuit
(d) a multimeter cannot be used to test for continuity

2. Which of the following results in loss of electrical energy from the circuit?

(a) Susceptance (b) Inductive reactance (c) Capacitive reactance (d) Resistance

3. The letters MTR on the drawing would indicate ____.

(a) middle of run (b) manufacture (c) motor (d) motor-tie-resistor

4. Ohm's Law cannot be applied for which of the following?

(a) To determine the voltage drop through a resistance.
(b) To determine the impedance of complete motor circuits.
(c) To determine the current produced by a voltage impressed on a resistance.
(d) To determine the voltage required to provide a specific current through a resistance.

5. If the scale is 1/8" = 1' and the conduit run measures 5 3/4" on the drawing, how many lengths of 10' conduit is needed?

(a) 3 (b) 4 (c) 5 (d) 6

6. A newly installed DC motor is running in the wrong direction. What change is necessary to correct this problem?

(a) Shift the brushes toward the direction of rotation.
(b) Interchange the connections of the field winding.
(c) Revolve the pole pieces 180° on their own axis.
(d) Reverse the line leads.

7. On which side of the door should switches be located?

(a) hinged side (b) opposite of hinged side (c) back side (d) door jam vertical

ANSWERS --- CLOSED BOOK (or no book) EXAM QUESTIONS - Practice

1. **(c) insert the test leads and connect the test leads in series with the circuit after de-energizing the circuit.** General knowledge

2. **(d) Resistance** AEH pg. 1.27, sec 50 (AEH is American Electricians Handbook)

3. **(c) motor** Plans review book

4. **(b) To determine the impedance of complete motor circuits.** AEH pg. 1.46 sec 107

5. **(c) 5** Solution: 5.75/.125 = 46'/10' = 4.6 or 5 lengths

6. **(b) Interchange the connections of the field winding.** AEH pg. 7.29 sec. 55

7. **(b) opposite of hinged side** General knowledge

PREPARING FOR AN OPEN BOOK EXAM

Part II Open Book

Most applicants agree this is the most difficult part of an electrical exam. Time becomes such an important factor. 50 open book questions are to be answered in two hours on the Journeyman exam.

Part II is a test of your knowledge and use of the National Electrical Code. 86% of the open book Journeyman questions are from the Code book.

Your score on the open book exam depends on how familiar you are with the Code book. Most exam applicants run out of time and are not able to find all the answers to the questions within the limited time.

Journeyman Exam
50 Questions
2 Hour Time Limit

That averages to
2.4 minutes
per question

The key to an open book exam is not to spend too much time on one question. If the question does not contain a key word that you can find in the index, **skip this question**, and continue to the next question. If you spend 3 minutes, 5 minutes, 6 minutes on a question and never find the answer you are eating into the time that should be used for the answers you can find.

In general there are usually 8 to 10 really difficult questions on an exam. The remaining questions after proper preparation you will be able to find within the alotted time. Skip these 8 or 10 as you recognize them and move on finding the other answers. If you answer 40 questions correctly out of a total of 50 questions your score would be 80%! That's better than in some cases where the applicant hasn't even answered 20 questions and time has ran out. You **can't** spend 5 or 6 minutes on a question. Never leave a question unanswered, unanswered is counted wrong. Always select a multiple choice answer before time runs out.

Proper preparation is so important in passing an open book exam. Don't be guilty of reading a question and feeling, "I know the answer so I won't bother looking in the Code book". The following pages will prove how this can be a big mistake. I teach by being properly prepared with how to find your way around in the Code book. You'll be able to look up all the answers within the time limit.

The difficulty occurs when you say Code book. Most applicants taking an exam are not familiar enough with the Code book and it's easy to understand why only 25 out of 100 pass an electrical exam.

JOURNEYMAN OPEN BOOK EXAM 50 QUESTIONS

Shown below are how the 50 questions are divided up.

- 14-16 questions are asked from Chapter 2 which represents 30% of the exam
- 14-16 questions are asked from Chapter 3 which represents 30% of the exam
- 9-11 questions are asked from Chapter 4 which represents 20% of the exam
- 4-6 questions are asked from theory which represents 10% of the exam
- 1-3 questions are asked on tools which represents 4% of the exam
- 1-3 questions are asked from Articles 100 and 110 which represents 4% of the exam
- 0-2 questions are asked from Articles 600 and 680 which represents 2% of the exam

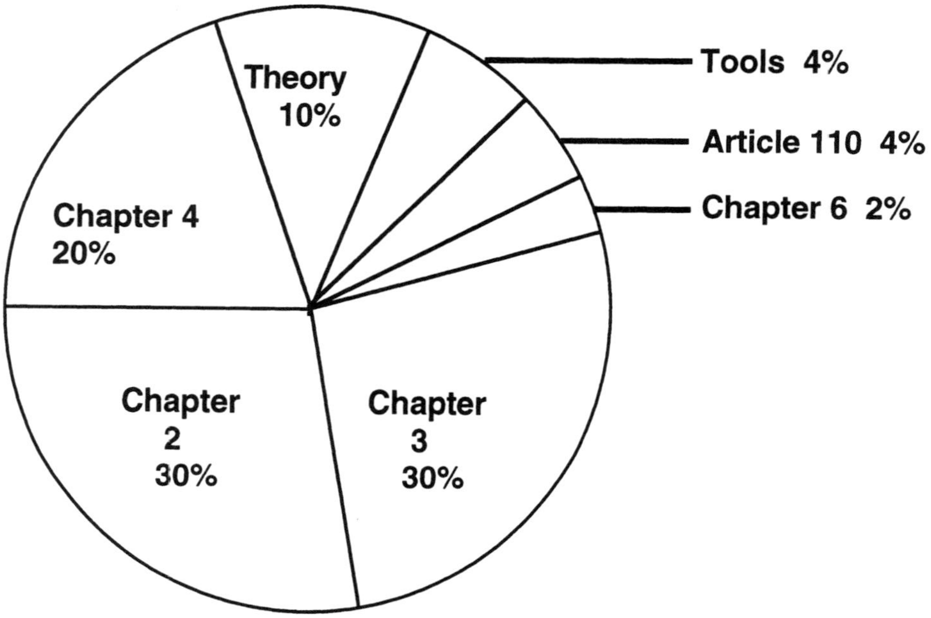

The best reference book for locating words in the Code book is "The Key Word Index". This book contains every word in the Code book with section number and page number. Now you can find what you're looking for in seconds! The "Key Word Index" is even pre-drilled with five holes so it can be added to the looseleaf Code book with ease. Now you'll be able to show them out on the job where it says that in the Code book. Try it once and you'll never be without it.

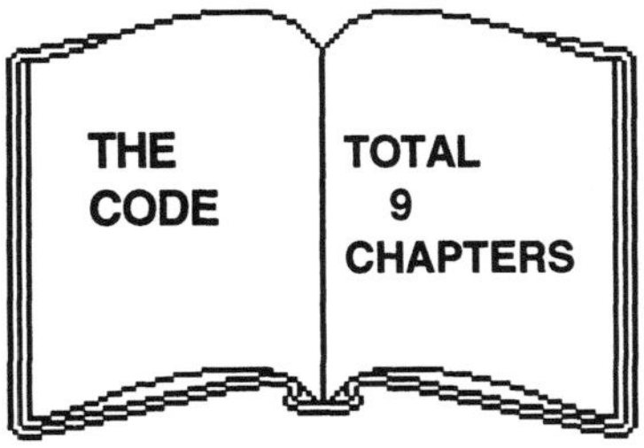

The most difficult task in preparing for the electrical exam is trying to "study" the Code.

The Code book is divided into nine chapters and then divided into articles, parts and sections.

The "meat" of the Code is the first four chapters. General wiring, grounding, services, motors, etc.

Chapters 5 through 9 are for special applications.

The following is an example of how the Code is divided: Flexible cords are not permitted as a substitute for fixed wiring of a structure per 400.8 of the Code.

The 400.8 is broken down to the 4 indicates Chapter 4.

The 400 is Article 400.

Article 400 is divided into three parts: Part A. General, Part B. Construction Specifications, Part C. Portable Cables over 600 Volts, Nominal.

400.8, the 8 is the section number.

Chapter 4, Article 400, Section 8 which is located in Part A of Article 400.

The latter part of each Article will contain the **over 600 volts (high-voltage) section**.

Example: The definition of a fuse is located in the **over 600 volts** Part B of Definitions Article 100 Chapter 1. Article 100 Definitions is listed in alphabetical order but fuse is not listed in Part A. Following the last Definition in Part A is the word wet location. Part B over 600 volts starts after wet location. Fuse is defined in Part B of Definitions not Part A.

It is very helpful as we try to master the Code book to know how it is laid out in Chapters, Articles, Parts and Sections.

After completing these practice exams turn to the answer sheets in the back of this book and grade yourself. 75% is passing.

To find your percentage simply divide the number of correct answers by the number of questions. Example: 38 correct answers divided by 50 questions would equal 76%.

I teach in my Electrical Code Classes that the key to the exam is that the student must first understand the question, which requires **careful reading of each word.**

As practice for the Open Book Exam Part II, try to find the correct answers for the following seven questions on the next page. Time limit for the 7 questions: 16 to 20 minutes. **GOOD LUCK!!**

After answering the seven open book questions turn to the next page for the answers with references and the index key format.

I chose seven of the more difficult questions from the Journeyman exam so the student will realize how important **TIME** is in this part of the exam. Again, the key is don't spend too much time on a difficult to locate question, **skip** over it and move on, it won't fail you!

This workbook was designed to help you with this difficult area of the exam. Some students purchase a Code book just prior to taking an exam, and as you can see after completing this workbook, you are expected to be an expert in finding references in the Code book. How do you study a Code book for an exam? This workbook is the best way as it **forces** you to find the answers in the time limit. Write down your score and time spent on each exam in this workbook and notice your improvement as you work the latter exams.

To pass the exam is very simple, it takes work! Like with anything in life, you get out of it what you put into it. The more time you spend preparing for the exam the easier it will become.

As you work these exams and grade yourself, hi-lite the answers with a marking pen in your Code book, or better yet purchase the "Ultimate Code Book" which has the complete package for taking an electrical exam.

The 2005 Ultimate Code
Item #620

Will Rogers once said, "You can't come back from someplace you've never been." This book will take you there.

OPEN BOOK PRACTICE 7 QUESTIONS TIME LIMIT: 16 MINUTES

1. Conductors supplying two or more motors shall have an ampacity equal to the sum of the full-load current rating of all the motors plus ____ % of the highest rated motor in the group.

(a) 25 (b) 80 (c) 100 (d) 125

2. In a residence, no point along the floor line in any wall space may be more than ____ feet from an outlet.

(a) 6 (b) 6 1/2 (c) 12 (d) 10

3. Which of the following statements about a #2 THHN cu conductor is correct?

(a) Its maximum operating temperature is 90° C
(b) It has a nylon insulation
(c) Its area is .067 square inches
(d) It has a DC resistance of .319 ohms per m/ft from Table 8

4. The maximum permissible open circuit voltage of electric-discharge lighting equipment used in a dwelling occupancy is ____ volts.

(a) 1000 (b) 120 (c) 240 (d) 50

5. Voltage shall not exceed 600 volts between conductors on branch circuits supplying only ballasts for electric-discharge lamps in tunnels with a height of not less than ____ feet.

(a) 12 (b) 15 (c) 18 (d) 22

6. Noninsulated busbars will have a minimum space of ____ inches between the bottom of enclosure and busbar.

(a) 6 (b) 8 (c) 10 (d) 12

7. If made up with threadless couplings, a 1" rigid metal conduit shall be supported at least every ____ feet.

(a) 6 (b) 8 (c) 10 (d) 12

ANSWERS TO OPEN BOOK PRACTICE   KEY TO USING NEC INDEX

1. **(a)** 430.24 Section 430.24 reads: Shall have of an ampacity not less than 125% of the full-load current rating of the highest rated motor plus the sum of the full-load current ratings of all other motors in the group. Question #1 reads: The sum of the full-load current rating of **ALL** the motors plus ____% of the highest rated motor. The highest rated motor was already included at 100% in the question in **ALL** motors, so you need to add 25% more to make it 125%.

2. **(a)** 210.52(A1). The key is to read the exact wording in the question. We space outlets 12' apart in a residence, but section 210.52(A1) states that **no point along the floor line** in any wall space is more than 6' from an outlet.

3. **(a)** Table 310.13. The key is THH**N**, the **N** is a nylon **covering**, not insulation. 0.067 square inches is from Table 8 which is for **bare** conductors. 0.319 ohms per m/ft from Table 8 is for **aluminum** not copper.

4. **(a)** 410.80(B). The key is to check the NEC index for "Electric discharge lighting", check each listing and 410-Q will lead you to the answer in article 410.80(A).

5. **(c)** 210.6(D1b). "Ballasts" and "Electric-discharge lamps" are listed in the index but are no help. "Tunnel" is not even listed in the index. The key, "lighting fixture voltages" from the index will lead you to article 210.6 and the answer.

6. **(c)** Table 408.5. "Noninsulated" is not listed in the index. "Busbars" is listed, but of no help. The key, think of where you would find a busbar located, in an enclosure, a **panelboard**. Article 408 will lead you to the answer in T.408.5.

7. **(c)** 344.30. The key word is **"threadless"** couplings. Section 344.30(B1) states a conduit shall be supported at least every 10'. 344.30(B2) states if made up with **threaded** couplings, you can use Table 344.30(B2) for supports.

When looking at the answer choices and then reading the Code book REMEMBER: 12" is 1 foot, 60 seconds is one minute, twelve is 12, 7 1/2' is 7' 6", 33% is one third, 6" is Trade size 6, 18" is 1 1/2', 2 1/2' is 30", 54" is 4 1/2', 2' is 24", and 6' 6" is 6 1/2'.

2005

CLOSED BOOK EXAM #1

50 QUESTIONS
TIME LIMIT - 1 HOURS

TIME SPENT ☐ **MINUTES**

SCORE ☐ **%**

MASTER CLOSED BOOK EXAM #1 **One Hour Time Limit**

1. Electroplating uses a ____ generator.

(a) series-wound (b) compound-wound (c) shunt-wound (d) separately-excited

2. The total opposition to the flow of alternating current is ____.

(a) resistance (b) impedance (c) induction (d) capacitance

3. 3 ø currents are generally out of phase by ____ degrees.

(a) 30 (b) 60 (c) 90 (d) 120

4. ____ is the ratio of output to input.

(a) Reluctance (b) Cosine (c) Efficiency (d) Square root

5. The voltage produced by electromagnetic induction is controlled by ____.

(a) the number of lines of flux cut per second (b) eddy currents
(c) the size of the magnet (d) the number of turns

6. Which of the following has the highest dielectric strength to electrical breakdown?

(a) thermoplastic (b) impregnated paper (c) rubber (d) woven cloth

7. The greatest voltage drop in a circuit will occur when the ____ the current flow through that part of the circuit.

(a) greater (b) slower (c) faster (b) lower

8. ____ results in loss of electrical energy from the circuit.

(a) Resistance (b) Reluctance (c) Susceptance (d) Admittance

9. Soft iron is most suitable for use in a ____.

(a) natural magnet (b) permanent magnet (c) magneto (d) temporary magnet

10. As the temperature increases, the resistance of most conductors also increases, except ____.

(a) silver (b) brass (c) carbon (d) zinc

11. Of the six ways of producing emf, which method is used the least?

(a) pressure (b) solar (c) chemical action (d) friction

12. Resistance in the power formula equals ____.

(a) E x I (b) E^2/W (c) E^2I (d) I^2/W

13. What is the formula to find watt hours?

(a) E x T x 1000 (b) E x I x T (c) I x E x T/1000 (d) E x T x ø/1000

14. In a series circuit when the voltage remains constant and the resistance increases, the current ____.

(a) increases (b) decreases (c) remains the same (d) increases by the square

15. Other factors remaining the same, the effect on the current flow in the circuit would cause the current to ____if the applied voltage was doubled.

(a) double (b) divide by 2 (c) remain the same (d) increase 4 times

16. A mil is what part of an inch?

(a) 1/10 (b) 1/100 (c) 1/1000 (d) 1/10000

17. A #12 wire has a diameter of 80.81 mils. The cma would be ____.

(a) 4110 (b) 5630 (c) 4374 (d) 6530

18. A substance whose molecules consist of the same kind of atoms is called ____.

(a) proton (b) valence (c) element (d) compound

19. The electrons in the last orbit of an atom are called ____ electrons.

(a) bound (b) free (c) valence (d) atomic

20. A length of wire has a resistance of 6 ohms. The resistance of a wire of the same material three times as long and twice the csa will be ____ ohms.

(a) 36 (b) 12 (c) 9 (d) 1

21. The hot resistance of a 100 watt incandescent bulb is about ____ times its cold resistance.

(a) 10 (b) 2 (c) 50 (d) 100

22. The purpose of load in an electrical circuit is to ____.

(a) utilize electrical energy (b) increase the current (c) decrease the current (d) none of these

23. A device for making, breaking, or changing connections in a circuit under load is a ____.

(a) inductor (b) growler (c) relay (d) switch

24. What relationship determines the efficiency of electrical equipment?

(a) The power input divided by the output
(b) The volt-amps x the wattage
(c) The va divided by the pf
(d) The power output divided by the input

25. The conductance of a conductor is the ease in which current flows through it. It is measured in ____.

(a) teslas (b) henrys (c) mhos (d) vars

26. In which of the following would a rheostat most likely not be used?

(a) transformer (b) motor (c) generator (d) motor-generator set

27. Nichrome wire having a resistance of 200Ω per 1000 feet is to be used for a heater that requires a total resistance of 10Ω. The length of wire required would be____ feet.

(a) 10 (b) 25 (c) 30 (d) 50

28. A fluorescent light that blinks "on" and "off" repeatly may in time ___.

(a) cause the fuse to blow
(b) cause the switch to wear out
(c) cause the wire to melt
(d) result in damage to the ballast

29. Electrical appliances are connected in parallel because it ____.

(a) makes the operation of appliances independent of each other
(b) results in reduced power loss
(c) is a simple circuit
(d) draws less current

30. If 18 resistances each 36 Ω are connected in parallel, the total resistance would be ____ Ω.

(a) 648 (b) 324 (c) 9 (d) 2

31. A negatively charged body has ____.

(a) excess of electrons (b) excess of neutrons (c) deficit of electrons (d) deficit of neutrons

32. A capacitor stores ____.

(a) voltage (b) power (c) current (d) charge

33. Electrical appliances are not connected in series because ____.

(a) series circuits are complicated
(b) appliances have different current ratings
(c) the power loss is too great
(d) the voltage is the same

34. A capacitor opposes ____.

(a) both a change in voltage and current
(b) change in current
(c) change in voltage
(d) none of these

35. Permanent magnets use ____ as the magnetic material.

(a) nickel (b) iron (c) hardened steel (d) soft steel

36. The armature current drawn by any DC motor is proportional to the ____.

(a) motor speed (b) voltage applied (c) flux required (d) torque applied

37. A component having no continuity would have ____ resistance.

(a) high (b) low (c) infinite (d) none of these

38. Basically all electric motors operate on the principle of repulsion or ____.

(a) magnetism (b) induction (c) resistance (d) capacitance

39. Generally the maximum ambient temperature for operating Class B motors is ____.

(a) over 175°C (b) 40°C (c) 60°C (d) 75°C

40. The change in direction of light passing from one area to another of different density is called ____.

(a) illusion (b) refraction (c) reflection (d) diffision

41. An electrician in the industry would first check the ____ to correct a low power factor.

(a) resistance (b) hysteresis (c) inductive load (d) reluctance

42. Tinning rubber insulated twisted cable is done to ____.

(a) make the strands stronger
(b) prevent chemical reactions between the copper and the rubber
(c) increase the resistance
(d) meet NEMA requirements

43. Inductance in a circuit ____.

(a) delays the change in current (b) prevents current from changing
(c) causes power loss (d) causes current to lead voltage

44. The AC system is preferred to the DC system because ____.

(a) DC voltage cannot be used for domestic appliances
(b) DC motors do not have speed control
(c) AC voltages can be easily changed in magnitude
(d) high-voltage AC transmission is less efficient

45. The breakdown voltage of an insulation depends upon ____ value of AC voltage.

(a) r.m.s. (b) effective (c) peak (d) 1.732 of peak

46. As the power factor of a circuit is increased ____.

(a) reactive power is decreased (b) active power is decreased
(c) reactive power is increased (d) both active and reactive power are increased

47. For the same rating, the size of a 3ø motor will be ____ a single phase motor.

(a) more than that of (b) same as that of (c) less than that of (d) none of these

48. Carbon brushes are used on a commutator because ____.

(a) they lubricate and polish (b) carbon is cheap (c) resistance is decreased (d) none of these

49. DC series motors are used in applications where ____ is required.

(a) constant speed (b) high starting torque (c) low no-load speed (d) none of these

50. The voltage per turn of the primary of a transformer is ____ the voltage per turn of the secondary.

(a) more than (b) the same as (c) less than (d) none of these

2005

CLOSED BOOK EXAM #2

50 QUESTIONS
TIME LIMIT - 1 HOURS

TIME SPENT ☐ **MINUTES**

SCORE ☐ **%**

MASTER CLOSED BOOK EXAM #2 One Hour Time Limit

1. A high spot temperature in a corroded electrical connection is caused by a(an) _____.

(a) increase in the flow of current through the connection
(b) decrease in the voltage drop across the connection
(c) increase in the voltage drop across the connection
(d) decrease in the effective resistance of the connection

2. A hickey is ____.

(a) a tool used to bend small sizes of rigid conduit **(b) a part of a conduit**
(c) not used in the electrical trade **(d) used only by a plumber**

3. The lubricant used for a motor sleeve bearing would be ____.

(a) vaseline (b) grease (c) oil (d) graphite

4. The resistance of a circuit may vary due to ____.

(a) a loose connection (b) change in voltage (c) change in current (d) induction

5.The service disconnecting means for each set of service-entrance conductors shall consist of not more than ____ switches.

(a) 6 (b) 8 (c) 10 (d) 12

6. A light bulb usually contains ____.

(a) air (b) neon (c) H2O (d) either a vacuum or gas

7. A ____ is an enclosure designed either for surface or flush mounting and provided with a frame, mat, or trim in which a swinging door or doors are or may be hung.

(a) cabinet (b) panelboard (c) cutout box (d) switchboard

8. Which of the following is a voltmeter?

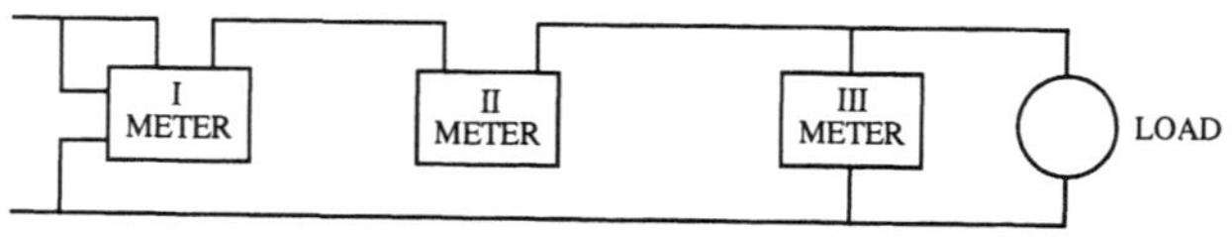

(a) I only (b) II only (c) III only (d) I, II or III

TH

9. Of the following ___ is a **false** statement.

(a) The term kilowatt indicates the measure of power which is all available for work.
(b) The term kilovolt-amperes indicate the apparent power made up of an energy component and a wattless or induction component.
(c) In an industrial plant, low power factor is usually due to underloaded induction motors.
(d) The power factor of a motor is much greater at partial loads than at full load.

10. A requirement of service that demands operation for alternate intervals of (1) load and no load; or (2) load and rest; or (3) load, no load, and rest is called ____ duty.

(a) variable (b) intermittent (c) short-time (d) periodic

11. An outlet where one or more receptacles are installed is called a ____.

(a) multi-outlet assembly (b) receptacle outlet (c) duplex outlet (d) tri-plex outlet

12. A conductor encased within material of composition or thickness not recognized by the Code is a ____ conductor.

(a) coated (b) semi (c) covered (d) fiber optic

13. Cooling of electrical equipment within enclosures is ____.

(a) the responsibility of the equipment manufacturer (b) not covered by the Code
(c) covered by the Code (d) not required

14. Approved is ____.

(a) listed and labeled equipment
(b) acceptable to the authority having jurisdiction
(c) tested and approved for the purpose by a qualified testing lab
(d) UL listed only

15. A system which will automatically furnish lighting and/or power to specified areas and/or equipment when there is a failure of the normal supply is known as a(n) ____ system.

(a) fail safe
(b) emergency
(c) alarm
(d) service safe

16. What chapter in the Code is Mobile Homes referred to?

(a) Chapter 3 (b) Chapter 5 (c) Chapter 8 (d) Chapter 9

17. Is it permissible to install direct current and alternating current conductors in one pull box?

(a) Yes, if insulated for the maximum voltage of any conductor.
(b) No, never.
(c) Yes, if ampacity is the same for both conductors.
(d) Yes, in dry places.

18. The maximum size fuse to be used in a branch circuit containing no motors depends on the _____.

(a) load (b) wire size (c) voltage (d) switch size

19. The type letter for moisture-resistant thermoset is _____.

(a) RUH (b) THW (c) RHW (d) MHR

20. On an Edison 3-wire system, some incandescent lamps are observed to be brighter, and others to be dimmer, than normal. What is the most likely trouble?

(a) weak breaker (b) loose fuse (c) poor neutral connection (d) too much voltage drop

21. A current-limiting overcurrent protective device is a device which will _____ the current flowing in the faulted circuit.

(a) reduce (b) increase (c) maintain (d) none of these

22. The conductors and equipment for delivering energy from the electricity supply system to the wiring system of the premises served is called the _____.

(a) primary (b) distribution (c) main supply feeder (d) service

23. Concealed is _____.

(a) not readily visible
(b) made inaccessible by the structure or finish of the building
(c) surrounded by walls
(d) attached to the surface

24. Continuous load is _____.

(a) a load where the maximum current is expected to continue for three hours or more
(b) a load where the maximum current is expected to continue for one hour or more
(c) intermittent operation in which the load conditions are regularly recurrent
(d) operation at a substantially constant load for an indefinitely long time

25. If a live conductor is contacted accidentally, the severity of the electrical shock is determined primarily by ____.

(a) whether the current is AC or DC **(b) the current in the conductor**
(c) the size of the conductor **(d) the contact resistance**

26. Electron flow produced by means of applying a pressure to a material is called ____.

(a) thermoelectricity (b) piezoelectricity (c) electrochemistry (d) photo conduction

27. A dual voltage motor will run more efficiently ____.

(a) at the lower voltage **(b) at the higher voltage**
(c) the same at either voltage **(d) none of these**

28. The term anode refers to ____.

(a) capacitor (b) dynamo (c) rectifier (d) inductor

29. A box contains one grounded and three ungrounded conductors, from one ungrounded to the grounded conductor 208 volts is measured, the other two ungrounded measure 120 volts to the grounded conductor, the system is ____.

(a) delta (b) wye 3-wire (c) wye 4-wire (d) open delta 4-wire

30. The voltage drop in a line can be decreased by ____.

I. increasing the resistance II. increasing the current
III. decreasing the load IV. increasing the wire size

(a) III only (b) I and III only (c) III and IV only (d) I, III and IV only

31. A 10 ohm resistance carrying 10 amperes of current uses ____ watts of power.

(a) 100 (b) 20 (c) 500 (d) 1000

32. When using a #12-2 with ground, the ground ____ carry current under normal operation.

(a) will (b) will not (c) will sometimes (d) none of these

33. A transformer is more efficiently utilized when the load has a ____ power factor.

(a) low (b) medium (c) average (d) high

34. When re-routing conduit, it may be necessary to increase the wire size if the distance is considerably greater, in order ____.

(a) to account for current drop
(b) to allow for possible resistance drop
(c) to compensate for voltage drop
(d) to increase the mechanical strength

35.To handle a three-phase unbalanced system, balance the system by making all loads equal to the ____ single phase load.

(a) smallest (b) average (c) largest (d) unbalanced

36. The connection between the grounded circuit conductor and the equipment grounding conductor at the service is called the ____.

(a) equipment bonding jumper
(b) main bonding jumper
(c) circuit bonding jumper
(d) electrode bonding jumper

37. Connections of conductors to terminal parts must ensure ____.

(a) a good connection without damaging the conductors
(b) proper torque values
(c) proper bonding
(d) proper crimping pressure

38. To determine directly whether all finished wire installations possess resistance between conductors, and between conductors and ground, use ____.

(a) set screws (b) shields (c) clamps (d) a megger

39. Most electric power tools, such as electric drills, come with a third conductor in the power lead which is used to connect the case of the tool to a grounded part of the electric outlet. The purpose for having this extra conductor is to ____.

(a) protect the user of the tool should the winding break down to the case.
(b) eliminate sparking between the tool and the material being worked upon.
(c) provide for continued operation of the tool should the regular grounded line-wire open.
(d) prevent accumulation of a static charge on the case.

40. The purpose of having a rheostat in the field of a DC shunt motor is to ____.

(a) control the speed of the motor
(b) minimize the starting current
(c) limit the field current to a safe value
(d) reduce sparking at the brushes

41. You are to check the power factor of a load, you cannot get a power factor meter, you would use ____.

(a) a wattmeter
(b) a voltmeter and an ammeter
(c) a kilo-watt hour meter
(d) an ammeter, a wattmeter and a voltmeter

42. The factor which will have the **least** effect on the voltage at the most distant point from the source of supply for a two wire circuit is ____.

(a) whether the supply is 25 cycle or 60 cycle AC
(b) the gauge of the circuit wires
(c) the amount of load on the circuit
(d) the length of the circuit

43. The load side is usually wired to the blades of a knife switch to ____.

(a) prevent blowing the fuse when opening the switch.
(b) make the blades dead when the switch is opened.
(c) prevent arcing when the switch is opened.
(d) allow changing of fuses without opening the switch.

44. High-voltage cable which is to be installed in underground ducts is generally protected with a ____.

(a) copper outer jacket (b) lead sheath (c) steel wire armor (d) tarred jute covering

45. A conduit coupling is sometimes tightened by using a strap wrench rather than a Stillson wrench, the strap wrench is used when it is important to avoid ____.

(a) crushing the conduit
(b) bending the conduit
(c) stripping the threads
(d) damaging the outside finish

46. The life of insulation used in electrical installations is directly affected by heat. Of the following, the electrical insulation which can **least** withstand heat is ____.

(a) mica (b) rubber (c) fiberglass (d) enamel

47. With respect to a common light bulb, it is correct to state that the ____.

(a) circuit voltage has no effect on the life of the bulb
(b) base has a left hand thread
(c) filament is made of carbon
(d) lower wattage bulb has the higher resistance

48. One identifying feature of a squirrel-cage induction motor is that it has no ____.

(a) air gap
(b) commutator or slip rings
(c) iron core in the rotating part
(d) windings on the stationary part

49. What is the voltage between points **Y** and **Z** ?

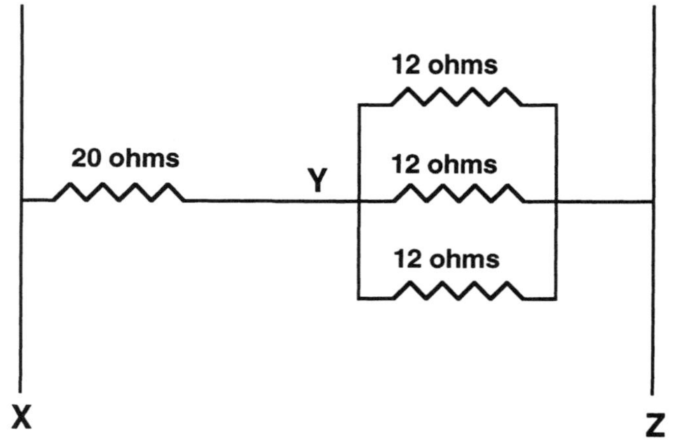

EACH OF THE 12 OHM LOADS IS 2 AMPERES

(a) 72 volts (b) 120 volts (c) 24 volts (d) 144 volts

50. In the sketch above, what is the wattage of the 20 ohm load?

(a) 120 watts (b) 48 watts (c) 144 watts (d) 720 watts

2005

CLOSED BOOK EXAM #3

50 QUESTIONS
TIME LIMIT - 1 HOURS

TIME SPENT ☐ **MINUTES**

SCORE ☐ **%**

MASTER CLOSED BOOK EXAM #3 One Hour Time Limit

1. A grounding conductor installed over lightning cables for the purpose of interconnecting the system ground electrodes and providing lightning protection for the cables is called a/an _____.

(a) air terminal (b) ground anchor (c) counterpoise (d) grounding connector rod

2. A surge of undirectional polarity is called ____.

(a) corona (b) flashover (c) skin effect (d) impulse

3. The minimum temperature at which a given liquid gives off vapor in sufficient concentration to form an ignitable mixture is the ____.

(a) absolute temperature (b) kindling temperature (c) flash point (d) heat of fusion

4. Emergency power panel conductors supplying a building are tapped on ____.

(a) any subfed panel (b) any circuit breaker main
(c) any feeder circuit (d) the line side of the service

5. The grounding electrode shall be ____.

(a) copper (b) copper-clad aluminum (c) aluminum (d) any of these

6. A _____ is usually made on cables after installation.

(a) ampacity test (b) no-load test (c) copper loss test (d) insulation resistance test

7. When you increase the resistance in a circuit, the flow of electrons will ____.

(a) flow faster (b) be stopped (c) be decreased (d) be constant

8. An instrument used for observing voltage or current waveforms is the ____.

(a) synchroscope (b) varameter (c) multimeter (d) oscilloscope

9. The ideal internal resistance of an ammeter should be ____.

(a) infinity (b) zero (c) equal to the circuit resistance (d) higher than the circuit resistance

10. The power factor of an alternator is determined by its ____.

(a) speed (b) load (c) excitation (d) weight

11. With the same voltage applied, ____ ohms allows more current to flow.

(a) 0.50 (b) 5 (c) 50 (d) 500

12. With respect to the safety value of the insulation on electrical tools, it can be said properly that ____.

(a) it adequately insures the safety of the user
(b) its value is mainly to the untrained electrician helper
(c) the insulation provides very little real protection
(d) the insulation should not be used as the only protective means

13. A transformer has a turns ratio of 4/1. What is the secondary peak voltage if the primary has an applied voltage of 115?

(a) 40.66 (b) 81.3 (c) 162.6 (d) 28.75

14. If the power factor of a circuit is unity, its reactive power is ____.

(a) zero (b) a maximum (c) equal to I^2R (d) a negative quantity

15. A/an ____ has the least number of electrons.

(a) semiconductor (b) conductor (c) insulator (d) none of these

16. A ____ is used to maintain the strength of a magnetic field.

(a) keeper (b) air gap (c) container (d) gauser

17. If a transformer bank is using an open delta connection, ____ single-phase transformers are interconnected.

(a) three (b) only one (c) two (d) two primary and one secondary

18. Which of the following raceways is NOT permitted to be used in a hazardous location?

(a) Liquidtight flexible metal conduit (b) Rigid non-metallic conduit
(c) Rigid metal conduit (d) None of these

19. A contact connected in a control circuit that will ensure that a particular sequence of operation is followed is called a/an ____.

(a) seal-in (b) interlock (c) overlatch (d) bonding jumper

20. A ____ is a typical type of thermostat used in appliances with heating elements.

(a) solder pot (b) bimetallic (c) melting alloy (d) mercury

21. ____ will cause a magnetic contactor to chatter.

(a) Low resistance (b) An overload (c) Low voltage (d) High current

22. The load in an electrical circuit is used to ____.

(a) generate electrical energy (b) transmit electrical energy
(c) utilize the electrical energy (d) cause a voltage drop

23. If two equal resistances connected in series across a certain are now connected in parallel across the same supply, the power produced will be ____ that of the series connection.

(a) one-fourth (b) one-half (c) two times (d) four times

24. Which of the following possess four valence electrons?

(a) insulators (b) semiconductors (c) semi-insulators (d) conductors

25. The capacitor opposes any change in voltage across it by ____.

(a) acting as a short circuit at time equal to infinity
(b) acting as a short circuit at time equal to zero
(c) passing a voltage proportional to the rate of change of current
(d) passing a current proportional to the rate of change of voltage

26. The force which sets up or tends to set up magnetic flux in a magnet is ____ force.

(a) electromotive (b) magnetomotive (c) dynamic (d) potential

27. During the short circuit test on a transformer, ____ is shorted.

(a) both sides (b) either side (c) low side (d) high side

28. In a given circuit, when the power factor is unity, the reactive voltamps are ____.

(a) zero (b) maximum (c) equal to the current (d) equal to the apparent power

29. A diagram showing the physical location of coils, contacts, motors and the like in their actual positions would be a ____ diagram.

(a) schematic (b) power flow (c) ladder (d) wiring

30. The power factor of a certain circuit in which the voltage lags behind the current is 80%. To increase the power to 100%, it is necessary to add ____ to the circuit.

(a) resistance (b) capacitance (c) inductance (d) impedance

31. The reciprocal of capacitance is called ____.

(a) elastance (b) permeability (c) conductance (d) permittivity

32. A ____ is the simplest form of a motor controller.

(a) relay (b) drum switch (c) toggle switch (d) magnetic contactor

33. In an open delta connected system ____.

(a) phase voltage is equal to line voltage
(b) phase voltage is less than the line voltage
(c) phase voltage is greater than line voltage
(d) phase voltage is zero

34. Battery cells are connected in parallel to increase ____.

(a) the voltage capacity of the cells
(b) the resistance capacity of the cells
(c) the current capacity of the cells
(d) the hysteresis

35. An operation in which the motor runs when the pushbutton is pressed and will stop when the pushbutton is released is called ____.

(a) plugging (b) clipping (c) reversing (d) inching

36. The period of an AC wave is ____.

(a) the ratio of amplitude to the frequency
(b) the same as the frequency
(c) the reciprocal of frequency
(d) not related to frequency

37. The kva rating of an ordinary two winding transformer is increased when connected as an autotransformer because ____.

(a) the secondary voltage is increased
(b) the secondary current is increased
(c) transformation ratio is increased
(d) energy is transferred both inductively and conductively

38. If the resistance in a circuit is increased, the flow of electrons will ____.

(a) be stopped (b) be decreased (c) be faster (d) be multiplied

39. The peak value of an AC sine wave is the same as the ____.

(a) effective value (b) rms value (c) instantaneous value (d) maximum value

40. A megger is conected to the ends of a motor winding. A low ohm reading indicates ____.

(a) open coil (b) continuity (c) loose coil (d) dirty coil

41. After the starting winding of a single-phase induction motor is disconnected from supply, it continues to run only on ____ winding.

(a) field (b) rotor (c) compensating (d) stator

42. An electrical temperature sensing device which is composed of a pair of different kinds of metal wires joined together in three complete loops is a ____.

(a) flowmeter (b) psychrometer (c) thermocouple (d) photoconductive cell

43. High voltage is used in long-distance power transmission because ____.

(a) it is easier to regulate than low voltage
(b) smaller transformers can be used
(c) the electromagnet fields are stronger
(d) the I^2R losses are lower

44. A DC generator develops ____ in its armature.

(a) alternating current (b) heat (c) inductance (d) direct current

45. When thermal overloads are used for the protection of a three-phase motor, their primary purpose is to protect the motor in case of ____.

(a) sustained overload
(b) reversal of phase sequence
(c) high voltage
(d) short circuit between lines

46. A disruptive discharge through the conductor insulation is called a ____.

(a) fault (b) short (c) breakdown (d) overload

47. An assembly of two pieces of insulating material provided with grooves for holding one or more conductors at a definite spacing from the surface wired over and from each other, and with holes for fastening in position is a ____.

(a) cleat (b) spool insulator (c) split knob (d) rosette

48. The disadvantage of a mechanical switch is that it ____.

(a) is costly (b) is operated mechanically (c) has high inertia (d) has a lower corona

49. Suppose the primary line voltage fluxuates between 105 volts and 125 volts on a 5/1 step down transformer. The maximum secondary peak voltage would be approximately ____ volts.

(a) 88 (b) 18 (c) 74 (d) 35

50. The force between two electrically charged objects is called ____.

(a) electroscopic force (b) magnetic force (c) eddy currents (d) electrostatic force

2005

CLOSED BOOK EXAM #4

50 QUESTIONS
TIME LIMIT - 1 HOURS

TIME SPENT ______ **MINUTES**

SCORE ______ **%**

MASTER CLOSED BOOK EXAM #4 — One Hour Time Limit

1. A 480/208/120v 3ø transformer is connected delta-wye. The secondary line voltage of the transformer is ____ in the secondary.

(a) less than the phase voltage **(b) equal to the phase voltage**
(c) greater than the phase voltage **(d) none of these**

2. The National Electrical Code rules that apply to electric welders are covered in Article ___.

(a) 520 (b) 600 (c) 620 (d) 630

3. ___ equipment or materials to which has been attached a label, symbol, or other identifying mark of an organization that is acceptable to the authority having jurisdiction and concerned with product evaluation that maintains periodic inspection of equipment or materials and indicates compliance with appropriate standards or performance.

(a) Listed (b) Labeled (c) Approved (d) Tested

4. What is the reason the Code requires each separate phase conductor to be located in the same raceway?

(a) to reduce inductive heat **(b) to improve workmanship**
(c) to provide identification of a circuit **(d) to assure balance resistance**

5. A three-phase delta-wye connected transformer, 480v primary, 208/120v secondary, the secondary line current is ____.

(a) equal to the secondary phase current **(b) greater than the secondary phase current**
(c) less than the secondary phase current **(d) 1.732 times the secondary phase current**

6. On a delta three-phase, 4-wire system, how many hot wires may use a common neutral?

(a) 2 (b) 3 (c) 4 (d) 6

7. The most common type of base used on standard incandescent light bulbs of 300 watts or less is ___.

(a) candelabra (b) admedium (c) mogul (d) medium

8. A ___ of a switch is that part of a switch which is used for the making or breaking of a connection and which is electrically insulated from other contact making or breaking parts.

(a) line terminal (b) pole (c) contact block (d) operating yoke

9. The point of attachment of service drop conductors to a building or other structure shall in no case be less than ___ feet.

(a) 10 (b) 12 (c) 15 (d) 18

10. Electrical controls for the ventilation system shall be arranged so that the air flow can be ____.

(a) reversed (b) continuous (c) maintained (d) constant

11. An isolated ground receptacle shall be identified by ___.

(a) CO/ALR orange marking on the face of the receptacle
(b) a metal faceplate not less than 0.030 inches in thickness
(c) a reset type test button the face of the receptacle
(d) an orange triangle located on the face of the receptacle

12. The speed of a DC motor can be controlled by ___.

(a) varying the armature current while the field current is held constant
(b) varying the field current while the armature current is held constant
(c) varying the armature voltage while the field voltage is held constant
(d) varying the field voltage while the armature voltage is held constant

13. If a three-phase, delta-wye transformer bank having a 480v primary and a 208/120v secondary, is considered to be 100% efficient, and to have resistive-type loads, the maximum kva of the load will be ____.

I. equal to the kva of the secondary of the transformer
II. equal to the primary kva of the transformer
III. considerably less than the kva of the transformer

(a) I only (b) II only (c) I and II only (d) I, II and III

14. Surrounded by a case, housing, fence, or walls that prevent persons from accidentally contacting energized parts is ___.

(a) enclosed (b) isolated (c) guarded (d) concealed

15. ___ means that equipment is not readily accessible to persons unless special means for access are used.

(a) Isolated (b) Guarded (c) Elevated (d) Concealed

16. ___ locations are those that are hazardous because of the presence of easily ignitable fibers or flyings.

(a) Class I (b) Class II (c) Class III (d) Class II, division II

17. The Code is divided into the introduction and nine chapters. Chapters 1, 2, 3, and 4 apply generally and the latter chapters supplement or modify the general rules except Chapter ___ which is independent of the other chapters except where specifically referenced therein.

(a) 6 (b) 7 (c) 8 (d) 9

18. A transformer would most likely have an efficiency of ____ percent.

(a) 60 (b) 70 (c) 80 (d) 90

19. Where lighting outlets are installed in interior stairways, there shall be a wall switch at each floor level to control the lighting outlet where the difference between floor levels is ___ risers or more.

(a) 1 (b) 3 (c) 5 (d) 6

20. The installation of wiring and equipment for fire alarm systems including all circuits controlled and powered by the fire alarm system is covered in Article ___.

(a) 695 (b) 710 (c) 725 (d) 760

21. A conducting material that is ___ charged will have an excess of electrons.

(a) neutrally (b) negatively (c) positively (d) none of these

22. When three single phase transformers are connected in such a way that the line current is equal to the phase current this transformer is ___ connected.

(a) delta (b) wye (c) delta zig zag (d) delta high leg

23. ___ losses in an iron core transformer occur when materials are used that retain a large part of their magnetization after the magnetizing force is removed. These materials are said to have a high permanence.

(a) Eddy current (b) Saturation (c) Copper (d) Hysteresis

24. The difference between a neutral and a grounded circuit conductor is ___.

(a) only a neutral will have equal potential to the ungrounded conductor
(b) only a neutrals outer covering is white or natural gray
(c) only a neutral carries unbalanced current
(d) there is no difference

25. A piece of electrical equipment that is designed to operate at alternate intervals of (1) load and no load; or (2) load and rest; or (3) load, no load, and rest is called ___ duty.

(a) short time (b) intermittent (c) periodic (d) varying

26. Insulated conductors of #___ or smaller, intended for use as grounded conductors of circuits shall have an outer identification of white or natural gray color.

(a) 6 (b) 4 (c) 3 (d) 1/0

27. ___ covers and plates shall be permitted with nonmetallic boxes.

I. Nonmetallic II. Metal

(a) I only (b) II only (c) either I or II (d) neither I or II

28. A light fixture to be installed in a wet location shall be ___.

I. installed so that water will not accumulate in wiring compartments
II. installed so that water will not enter the wiring compartment
III. marked suitable for damp locations

(a) I only (b) I and II only (c) II and III only (d) I, II, and III

29. The general requirements for communication circuits such as telephone, telegraph, outside wiring for fire alarm and burglar alarms is covered in Article ___ of the NEC.

(a) 720 (b) 725 (c) 760 (d) 800

30. The general requirements for all circuits and equipment operating at more than 600 volts nominal is mostly covered in Article ___.

(a) 695 (b) 490 (c) 725 (d) 760

31. Thermal protection is required by the NEC for ___.

I. fluorescent lighting fixture ballast installed indoors
II. fluorescent lighting fixture ballast installed outdoors
III. recessed incandescent fixtures installed indoors

(a) I and II only (b) I and III only (c) II and III only (d) I, II, and III

32. All conductors of the same circuit and, where used, the grounded conductor and all equipment grounding conductors shall be contained within the same ___.

I. cable II. raceway III. trench

(a) I only (b) II only (c) I and II only (d) I, II, and III

33. When a circuit breaker is in the OPEN position ____.

I. you have a short in the ungrounded conductor
II. you have a short in the grounded conductor

(a) I only (b) II only (c) either I or II (d) both I and II

34. An electrical outlet constructed so that moisture will not enter the enclosure is classified as being ____.

(a) waterproof (b) rainproof (c) watertight (d) weatherproof

35. You have an adjustable trip coil rated at 5 amps on a 200-amp switch.If you want the switch to trip at 120 amps, the trip coil should be set at ____.

(a) 2 amps (b) 3 amps (c) 4 amps (d) 5 amps

36. After cutting a conduit, to remove the rough edges on both ends, the conduit ends should be ____.

(a) sanded (b) shaped (c) burnished (d) ground

37. One should use an insulated basket grip to attach a pull line to the conductors because a bare steel grip should not be used because it will ___.

(a) react with the conductor
(b) stretch the conductor
(c) score the conduit during the pull
(d) reduce the coefficient of friction

38. The instrument used to indicate phase relation between current and voltage is the ____.

(a) megger (b) power factor meter (c) voltmeter (d) galvanometer

39. Reactance will cause the current in a circuit to vary only when ____.

(a) AC current flows **(b) DC current flows**
(c) there is no resistance in the circuit **(d) there is resistance in the circuit**

40. Fixtures supported by the framing members of suspended ceiling systems shall be securely fastened to the ceiling framing member by mechanical means. This would include all but which of the following?

(a) fixture wire (b) bolts or screws (c) rivets (d) clips identified for this use

41. With two conductors installed in a conduit, the conduit can be filled to ____ % of its cross section.

(a) 53 (b) 31 (c) 40 (d) 60

42. In a residence, no point along the floor line in any wall space may be more than ____ feet from an outlet.

(a) 6 (b) 6 1/2 (c) 12 (d) 10

43. Which of the following electrodes must be supplemented by an additional electrode?

(a) metal underground water pipe **(b) metal frame of a building**
(c) ground ring **(d) concrete encased**

44. Plaster, drywall or plasterboard surfaces that are broken or incomplete shall be repaired so there will be no gaps or open spaces greater than ____ inch at the edge of the fitting box.

(a) 1/16 (b) 1/8 (c) 3/16 (d) 1/4

45. When making electrical connections a torque screwdriver or wrench is required because ______.

I. many terminations and equipment are marked with a tightening torque
II. control circuit devices with screw-type pressure terminals used with #14 or smaller copper conductors shall be torqued to a minimum of 7 lb-in. unless identified for a different torque value
III. listed or labeled equipment shall be installed and used in accordance with any instructions included in the listing or labeling

(a) I only (b) I and III only (c) II only (d) I, II, and III

46. The SPST switch can control ____.

(a) one circuit
(b) two circuits, but not at the same time
(c) two circuits at the same time
(d) more than two circuits

47. In the "OFF" position, a good switch will have ____ resistance.

(a) high (b) moderate (c) zero (d) variable

48. In the circuit below which resistor will consume the most power?

(a) 2Ω (b) 4Ω (c) 6Ω (d) 8Ω

49. In the circuit below which resistor will consume the most power?

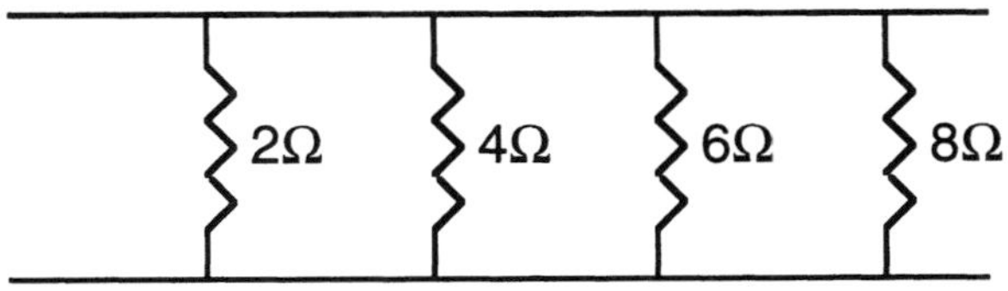

(a) 2Ω (b) 4Ω (c) 6Ω (d) 8Ω

50. The tool that is used to align vitrified tile conduit in multiple ducts is a ____.

(a) mandrel (b) manometer (c) growler (d) hickey

2005

CLOSED BOOK EXAM

#5

50 QUESTIONS
TIME LIMIT - 1 HOURS

TIME SPENT ☐ **MINUTES**

SCORE ☐ **%**

MASTER CLOSED BOOK EXAM #5 — One Hour Time Limit

1. Transformers are rated in kva instead of kw because _____.

(a) load power factor is often unknown **(b) total transformer loss depends on va**
(c) kw depends on the power factor **(d) none of these**

2. When two resistances are connected in parallel _____.

(a) the current through each must be the same
(b) the voltage across each must be the same
(c) each must have the same resistance value
(d) their combined resistance equals the sum of the individual values

3. The main disadvantage of a low power factor is that _____.

(a) heat generated is more than the desired amount
(b) current required for a given load power is higher
(c) more power is consumed by the load
(d) active power developed by a generator exceeds its rated output capacity

4. A _____ has the highest operating efficiency.

(a) converter (b) generator (c) transformer (d) motor

5. If a motor runs but fails to stop even if the stop button is pressed, the cause is probably _____.

(a) the overload contact did not operate **(b) the fuse has blown**
(c) the holding circuit interlock was welded **(d) all of these**

6. Overcurrent devices shall not be connected in series with any conductor that is _____.

(a) current carrying (b) intentionally grounded (c) stranded (d) above 250 kcmil

7. If the length and area of cross-section of a wire are doubled, then its resistance _____.

(a) remains the same (b) becomes doubled (c) becomes four times (d) is less

8. A good conductor material should have _____ valence electrons.

(a) 1 (b) 4 (c) 19 (d) 29

9. When the power factor of a circuit is zero the ____.

(a) impedance is minimum **(b) power absorbed is zero**
(c) power absorbed is minimum **(d) power absorbed is maximum**

10. A capacitor consists of ____.

(a) conductors only **(b) conductors separated by a dielectric**
(c) dielectric only **(d) dielectric separated by a conductor**

11. In a circuit breaker, the current which exists at the instant of contact separation is known as the ____ current.

(a) restriking (b) surge (c) recovery (d) interrupting

12. In motor controls, a maintaining contact is a ____ contact.

(a) delay-on (b) delay-off (c) normally closed (d) normally open

13. In an AC wave, 30 degrees of phase is ____ of a cycle.

(a) 1/4 (b) 1/3 (c) 1/2 (d) 1/12

14. The air space between the poles of a magnet is called ____.

(a) the air gap (b) the free zone (c) the free space (d) the vacuum

15. A repulsion-start induction-run single-phase motor runs as an induction only when ____.

(a) stator winding is reversed **(b) short-circuiter is disconnected**
(c) commutator segments are short-circuited **(d) brushes are shifted to neutral plane**

16. ____ is a disruptive discharge between electrodes of a measuring gap, voltage gap or protective device.

(a) Fire over (b) Flashover (c) Spark-over (d) Corona

17. At leading power factor, the armature flux of an alternator ___ the rotor flux.

(a) aids (b) opposes (c) distorts (d) does not effect

18. The connection between conductive or inductive metal objects in an element of a lightning protection system to accomplish continuity is ____.

(a) counterpoise (b) bonding (c) interlink (d) connectors

19. Overcurrent protection devices in emergency systems shall ____.

(a) not trip the main device (b) clear in steps (c) be coordinated (d) all of these

20. A ____ includes any switch or device normally used to start and stop a motor by making and breaking the motor circuit current.

(a) DPDT switch (b) rheostat (c) heater (d) controller

21. If there are no overcurrent protective devices rated 30 amps or less with neutral connection, the panelboard is classified as a ____.

(a) appliance panelboard (b) lighting panelboard (c) power panelboard (d) all of these

22. Overcurrent in transformers affect all of the following EXCEPT ____.

(a) life of the insulation (b) mechanical stresses
(c) rise in temperature (d) breather effectiveness

23. A relay which prevents holding in an electrically operated device, such as a circuit breaker, while an abnormal condition exists in the circuit is called ____.

(a) locking relay (b) trip-free (c) auxiliary relay (d) shunt relay

24. The essential condition for parallel operation of two single phase transformers is that they should have the same ____.

(a) polarity (b) percentage impedance (c) kva rating (d) voltage rating

25. A 10 amp fan with a PF of 80% is connected to a 230 volt source. The power in watts is ____.

(a) 2300 (b) 2175 (c) 1920 (d) 1840

26. Sparking between contacts can be reduced by ____.

(a) inserting a capacitor in series with the contacts (b) inserting a resistance in the line
(c) inserting a capacitor in parallel with the contacts (d) all of these

27. Autotransformers used to start large induction motors are frequently called a starting ____.

(a) winder (b) transformer (c) reactor (d) compensator

28. The reciprocal of resistance is ____.

(a) reluctance (b) susceptance (c) admittance (d) conductance

29. The switch with the fastest speed of operation is a/an ____ switch.

(a) hydraulic (b) mechanical (c) electromechanical (d) electronic

30. If you double the voltage in a circuit and cut the resistance in half, the current will become ____.

(a) the same as before (b) twice as great (c) half as great (d) four times as great

31. When a stationary three-phase induction motor is switched on with one phase disconnected ____.

(a) it will start very slowly
(b) it will jerk when started and make growling noise
(c) it is likely to burn out quickly unless disconnected immediately
(d) remaining intact fuses will be blown out due to heavy inrush current

32. The reason why alternating current can induce voltage is ____.

(a) it has a constant magnetic field (b) it has a stronger magnetic field than direct current
(c) it has a high peak value (d) it has a varying magnetic field

33. ____ is a term used to express the amount of electrical energy in an electrostatic field.

(a) Coulombs (b) Volts (c) Joules (d) Watts

34. Insulation resistance is measured with a/an ____.

(a) ohmmeter (b) megger (c) insulation meter (d) wein bridge

35. Inverse-time characteristics of a circuit breaker means ____.

(a) higher fault current, longer time needed to trip
(b) lower fault current, shorter time needed to trip
(c) higher fault current, shorter time needed to trip
(d) none of these

36. Which of the following is the best advantage of a DC motor over an AC motor?

(a) It has better speed control (b) It has a higher speed rating
(c) It is easier to reverse its speed (d) all of these

37. An electrical symbol represented by a rectangle with the letters PB inside is ____.

(a) battery panel (b) push button (c) pull box (d) power bend

38. The resistance of a component having no continuity is ____.

(a) infinite resistance (b) no or zero resistance (c) low resistance (d) all of these

39. A shunt for a milliammeter _____.

(a) extends the range and reduces the meter resistance
(b) extends the range and increases the meter resistance
(c) decreases the range and the meter resistance
(d) decreases the range but increases the meter resistance

40. The air space between poles of a magnet is called ____.

(a) vacuum (b) free space (c) null area (d) air gap

41. The basic unit of electric charge is the ____

(a) coulomb (b) ampere hour (c) farad (d) watt hour

42. A series circuit with a 8Ω resistance and a 4Ω resistance connected to a 12 volt battery will have a current flow of ____.

(a) 1.3 in the larger resistance (b) one ampere in both resistors
(c) 2.0 amps in both resistors (d) 3 amps in the smaller resistor

43. The property of magnetic materials of retaining magnetism after withdrawal of the magnetizing force is referred to as ____.

(a) resistivity (b) reluctivity (c) retentivity (d) reluctance

44. A device capable of drawing lightning discharge to it in preference to vulnerable parts of the protected area is a ____.

(a) ground terminal (b) lightning trap (c) air terminal (d) ground mat

45. A megger is an instrument to measure ____.

(a) very high resistance (b) very low resistance (c) inductance of a coil (d) insulation resistance

46. An open coil has ____.

(a) infinite resistance and inductance (b) zero resistance and infinite inductance
(c) zero resistance and inductance (d) infinite resistance and zero inductance

47. An electric motor in which the rotor and stator fields rotate with the same speed is called a/an ____ motor.

(a) universal (b) DC (c) synchronous (d) asynchronous

48. If the starting winding of a single-phase induction motor is left in the circuit, it will ____.

(a) run faster **(b) stop operating**
(c) draw excessive current and overheat **(d) spark at light loads**

49. The maximum load consumed or produced by a unit or group of units in a stated period of time is the ____ load.

(a) connected (b) continuous (c) average (d) peak

50. The permanent joining of metallic parts to form an electrically conductive path which will assure electrical continuity and the capacity to conduct safely any current likely to be imposed is ____.

(a) welding (b) splicing (c) molding (d) bonding

2005

CLOSED BOOK EXAM #6

50 QUESTIONS
TIME LIMIT - 1 HOURS

TIME SPENT ☐ **MINUTES**

SCORE ☐ **%**

MASTER CLOSED BOOK EXAM #6 One Hour Time Limit

1. A transformer is associated with ____ current.

(a) alternating (b) direct (c) neither alternating nor direct (d) either alternating or direct

2. Relay contacts are made of ____.

(a) copper (b) aluminum (c) silver (d) gold

3. A device that serves to govern, in some predetermined manner, the electric power delivered to the apparatus to which it is connected is called a ____.

(a) switch (b) control-circuit (c) feeder (d) controller

4. Conductors drawn from a copper-clad aluminum rod with the copper metallurgically bonded to an aluminum core. The copper forms a minimum of ____ percent of the csa.

(a) 10 (b) 20 (c) 40 (d) 70

5. Operation of equipment in excess of normal, full-load rating, or of a conductor in excess of rated ampacity which, when it persists for a sufficient length of time, would cause damage or dangerous overheating is called ____.

(a) a short-circuit (b) an overload (c) a ground-fault (d) induction

6. In an AC circuit a value of watts divided by a value of volt-amps, which of the following indicates close to unity?

(a) maximum current for the load (b) high power factor
(c) maximum voltage for the load (d) low power factor

7. A transformer core is made of ____.

I. copper II. steel III. aluminum

(a) I only (b) II only (c) III only (d) I, II or III

8. A qualifying term indicating that the circuit breaker can be set to trip at various values of current and/or time within a predetermined range is called ____.

(a) adjustable (b) instantaneous trip (c) setting (d) inverse time

9. A motor control circuit is the circuit of a control apparatus or system that carries the _____.

(a) load (b) power (c) signals (d) energy

10. It is customary to speak of the electromotive force as the ____of a circuit.

(a) voltage (b) rms (c) current (d) impedance

11. Which of the following is the symbol for a delta transformer connection?

(a) ø (b) Y (c) Ω (d) Δ

12. Flexible metal conduit shall be secured by an approved means at intervals not exceeding ____ feet and within 12" on each side of every outlet box, junction box, cabinet, or fitting.

(a) 3 (b) 4 (c) 4 1/2 (d) 6

13. The resistance of a 3.6 kw heater when operated from a 120 volt circuit is ____ ohms.

(a) 3 (b) 4 (c) 30 (d) 120

14. Materials containing numerous free electrons are ____.

(a) ferrous materials (b) good insulators
(c) good conductors (d) high numerally resistivity scales

15. An insulated conductor with a white or gray finish shall be permitted as _____ conductor where permanently reidentified to indicate its use, by painting or other effective means at its termination, and at each location where the conductor is visible and accessible.

I. grounded II. hot III. grounding IV. ungrounded

(a) I only (b) II and IV only (c) III only (d) II only

16. If the DC resistance of a #500 kcmil copper conductor 1000 feet in length is .0258Ω the DC resistance for a #1000 kcmil copper conductor 1000 feet in length would be _____ Ω.

(a) 0.0258 (b) 0.01515 (c) 0.0129 (d) 0.00643

17.

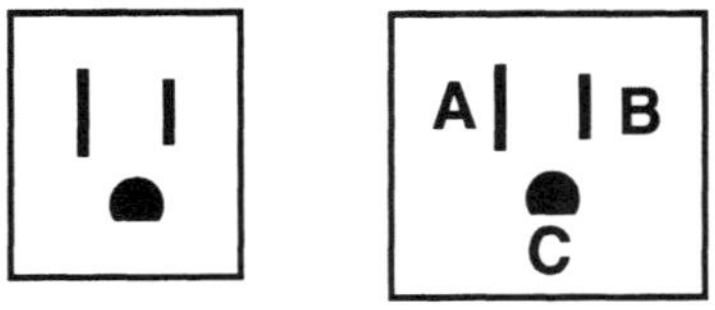

Which terminal on the receptacle is connected to the grounded conductor?

(a) A (b) B (c) C (d) none of these

18. A three-phase ungrounded 480v system in a plant has three ground detector lights. Two lamps are bright and "B" phase lamp is dark. This indicates that there is a ground fault on _____.

(a) A phase (b) B phase (c) C phase (d) any two phases have shorted

19. Metal cabinets used for lighting circuits are grounded in order to _____.

I. eliminate electrolysis
II. simplify wiring
III. limit the voltage to ground
IV. facilitate overcurrent device operation

(a) I only (b) I and II only (c) III only (d) III and IV only

20. Article 210 covers requirements for general purpose branch circuits. Article _____ provides rules for branch circuits that supply only motor loads.

(a) 460 (b) 470 (c) 445 (d) 430

21. The _____ is the current in amperes a conductor can carry continuously under the conditions of use without exceeding its temperature rating.

(a) load (b) demand (c) connected load (d) ampacity

22. The chain wrench can be used _____.

I. with one hand after the chain is around the conduit
II. in confined places and close to a wall
III. for many sizes of conduit

(a) I, II and III (b) I and II only (c) I and III only (d) II and III only

23. Nipples are short pieces of conduit ____.

(a) threaded at both ends **(b) threaded on one end**
(c) that may or may not be threaded **(d) that are never threaded**

24. Encased with a material or composition or thickness that is not recognized by the Code as electrical insulation is defined as a covered ____.

(a) cable (b) conduit (c) wire (d) conductor

25. An accessory such as a locknut, bushing, or other part of a wiring system that is intended primarily to perform a mechanical rather than an electrical function would be a _____.

(a) device (b) fitting (c) joint (d) guerney

26. Enclosures for overcurrent devices shall be mounted in a ____ position unless in individual instances this is shown to be impracticable and is installed in accordance with 240-81.

(a) horizontal (b) diagonal (c) open (d) vertical

27. An overcurrent trip unit of a circuit breaker shall be connected in series with each _____ .

(a) ungrounded conductor (b) grounded conductor (c) overcurrent device (d) transformer

28. Receptacle outlets in a residence shall be installed so that no point along the floor line in any wall space is ____, measured horizontally, from an outlet in that space.

(a) less than 6' (b) more than 6' (c) less than 12' (d) none of these

29. A ground is a conducting connection, whether ____ or accidental, between an electrical circuit or equipment and the earth, or to some conducting body that serves in place of the earth.

(a) identifying (b) intentional (c) conducted (d) none of these

30. Boxes used for ceiling outlets in all types of concealed wiring installed in buildings of all types of construction are ____.

(a) square (b) oblong (c) octagonal (d) devices

31. Grounding conductor shall not be connected by ____.

(a) exothermic welding (b) soldered fittings (c) listed pressure connectors (d) listed clamps

32. Article 665 Induction and Dielectric Heating covers the construction and installation of induction and dielectric heating equipment and accessories for ____ applications.

I. industrial II. scientific III. medical IV. dental

(a) I only (b) I and II only (c) I, II and III only (d) I, II, III and IV

33. You have 120 volts at the panel and 115 volts at the load. What is the percentage of voltage drop?

(a) 5% (b) 4.35% (c) 4.17% (d) 3%

34. The ampacity of a solid or stranded conductor of the same material is the same if it has the same ____.

(a) diameter (b) circumference (c) cross-sectional area (d) insulation

35. The____ is the service conductors between the terminals of the service equipment and a point usually outside the building, clear of building walls, where joined by tap or splice to the service drop.

I. service drop II. service entrance conductors, overhead system III. service equipment

(a) I only (b) II only (c) III only (d) I, II and III

36. If the circuit voltage is increased, all else remains the same, only the ____ will change.

(a) resistance (b) current (c) ampacity (d) conductivity

37. To ensure proper adhesion between wire, solder and a lug, the temperature of the ____ must be above the melting point when they are brought together in soldering.

I. wire II. solder III. lug

(a) I, II and III (b) I and II only (c) II and III only (d) II only

38. The conductor with the highest insulation temperature rating is ____.

(a) THWN (b) RH (c) RHH (d) THW

39. The current flow in this circuit would be ____ amps.

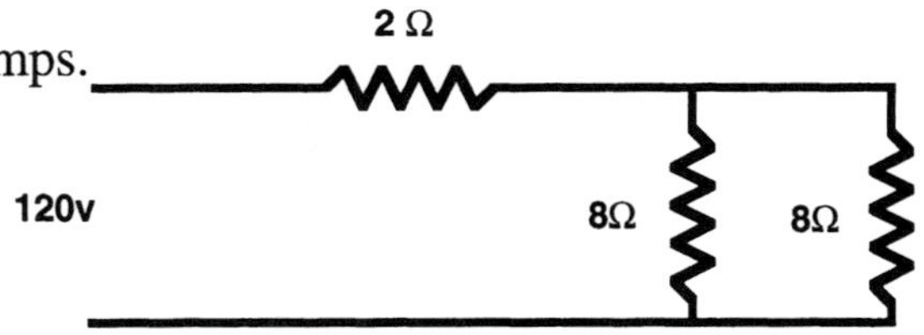

(a) 4 (b) 6 (c) 20 (d) 30

40. One should use an insulated basket grip to attach a pull line to the conductor. A bare steel grip should not be used because it will ____.

(a) react with the conductor **(b) stretch the conductor**
(c) score the conduit during the pull **(d) reduce the coefficient of friction**

41. The resistance of an electrical conductor increases with ____.

I. an increase in ambient temperature II. a decrease in csa III. an increase in conductor length

(a) I only (b) I and II only (c) III only (d) I, II and III

42. In a 6-pole machine, 360 electrical degrees is equal to ____ mechanical degrees.

(a) 60 (b) 90 (c) 120 (d) 180

43. The minimum cover requirement for rigid nonmetallic conduit without concrete encasement is ____ inches.

(a) 6 (b) 12 (c) 18 (d) 24

44. A nipple can be filled ____ percent of its total csa.

(a) 31 (b) 40 (c) 53 (d) 60

45. When calculating the cubic inch capacity required for a box, 3 equipment grounding conductors are counted as ____ conductor/s.

(a) 1 (b) 2 (c) 3 (d) not counted

46.  In a motor control wiring diagram this is the symbol for a ____.

(a) circuit breaker (b) overload interlock (c) thermal overload (d) bimetallic strip

47. How many receptacles are required in the hallway of a residence that is 15 feet in length?

(a) 1 (b) 2 (c) 3 (d) none are required

48. Which of the following shown below would calculate at 180 volt-amps?

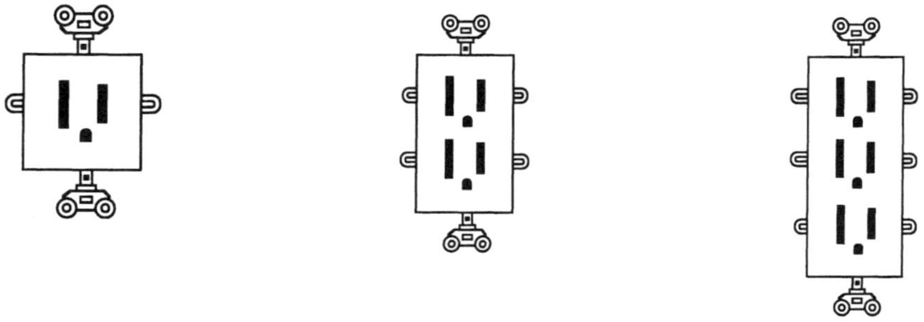

I. II. III.

(a) I only (b) I or II only (c) II only (d) I, II or III

49. Article ____ provides requirements for the installation of optical fibers in ducts and plenums.

(a) 820 (b) 770 (c) 710 (d) 545

50. A device that establishes a connection between two or more conductors or between one or more conductors and a terminal by means of mechanical pressure and without the use of solder is defined as a ____.

(a) shrink sleeve connector (b) boot connection (c) pressure connector (d) T-tap

2005

CLOSED BOOK EXAM

#7

50 QUESTIONS
TIME LIMIT - 1 HOURS

TIME SPENT ☐ **MINUTES**

SCORE ☐ **%**

MASTER CLOSED BOOK EXAM #7 One Hour Time Limit

1. Power factor correction _____.

(a) increases the reactive power taken by the load
(b) increases the active power drawn by the load
(c) employs series resonance
(d) does not affect the performance characteristics of the original load

2. At the same frequency, larger capacitance results in ____.

(a) less reactance (b) more reactance (c) the same reactance (d) none of these

3. A positive electric pole ____.

(a) has more electrons than the negative pole
(b) has an excess of electrons
(c) has fewer electrons than the negative pole
(d) has a deficiency of electrons

4. An actuator ____.

(a) converts light to heat
(b) converts an electrical input into action
(c) converts a non-electrical input to electrical output
(d) reduces the effects of noise and other disturbances on the measured quantity

5. Demagnetizing by applying an AC field and gradually reducing it to zero is called ____.

(a) hysteresis (b) gaussing (c) degaussing (d) damping

6. A ____ test is the preferred test used to check switchgear insulation.

(a) Varley loop (b) megohmmeter (c) AC high potential (d) DC high potential

7. The purpose of a breather on a transformer is to ____.

(a) provide cooling to the winding
(b) take insulating oil from conservator
(c) provide insulation to the winding
(d) extract moisture in the air

8. A modern type controller is called a ____.

(a) programmable logic counter
(b) programmable logic controller
(c) programmable laboratory controller
(d) programmable language controller

9. When current flows through a resistance, heating results. This condition called magnetorestriction is also known as ____.

(a) Joule (b) Edison (c) Tesla (d) Kelvin

10. An enclosure either above or below ground, with fire resistant walls, ceiling and floor exclusively built for unattended transformer and their auxiliaries is a transformer ____.

(a) vault (b) yard (c) compound (d) housing

11. The ____ method of starting a motor keeps the starting current at a minimum.

(a) star-delta (b) autotransformer (c) direct on line (d) across the line

12. The conductors after the final overcurrent device and before the load served are ____.

(a) branch circuit conductors (b) feeder conductors
(c) service conductors (d) control conductors

13. A disruptive discharge around or over the surface of a solid or liquid insulator is a ____.

(a) surge (b) corona (c) flashover (d) sparkover

14. The horizontal distance between two adjacent supporting points of a conductor is called the ____.

(a) sag (b) span (c) space (d) clearance

15. A transformer having 1000 primary turns is supplied by 250 volts. For a secondary voltage of 400 volts, the number of secondary turns is ____.

(a) 0.625 (b) 1.6 (c) 160 (d) 1600

16. Inserting an ammeter in a circuit will ____ the resistance of the circuit and therefore ____ the current in the circuit.

(a) reduce, increase (b) increase, reduce (c) increase, increase (d) decrease, decrease

17. An closed switch has a resistance of ____.

(a) infinity (b) zero (c) approximately 1000Ω (d) at least 300Ω

18. The magnetizing current drawn by transformers and induction motors is the cause of their ____ power.

(a) leading (b) lagging (c) zero (d) unity

19. A potentiometer is used to control the ____ of a circuit.

(a) voltage (b) current (c) resistance (d) wattage

20. A ____ is a combination of the arc discharge characteristics of a fluorescent lamp and the compact focusable shape of an incandescent lamp.

(a) tungsten halogen lamp (b) quartz lamp (c) sodium lamp (d) mercury lamp

21. A ____ is a device used for the purpose of minimizing irregularities in the flow of welding currents.

(a) reactor (b) grounding transformer (c) rheostat box (d) none of these

22. ____ is a term that means the motor will stop when there is a supply voltage failure and the motor will restart automatically when the supply voltage is restored.

(a) No voltage release (b) No voltage control (c) No voltage drop (d) No voltage hold

23. A 50 kva transformer has a primary voltage of 6600 volts and a secondary voltage of 250 volts. If it has 52 secondary turns, the number of primary turns would be approximately ____.

(a) 13,000 (b) 130 (c) 5 (d) 1373

24. In an AC circuit, the power dissipated as heat depends on the ____.

(a) resistance (b) impedance (c) capacitive reactance (d) inductive reactance

25. The loss of electrical energy in counter balancing the residual magnetism in each cycle is ____.

(a) hysteresis loss (b) eddy current loss (c) copper loss (d) leakage loss

26. The resistance of an insulator ____ when its temperature is increased.

(a) decreases (b) increases (c) varies (d) remains the same

27. An oscilloscope is usually used to measure ____ voltage.

(a) effective (b) average (c) rms (d) maximum

28. A 25 watt incandescent bulb rated at 120 volts and operated on a 120 volt line has burnt out and has to be replaced as soon as possible. There are several lamps available but not of the same rating. Which of the bulbs below should be used to approximate the power consumption of the burnt out bulb?

(a) 20w, 110v (b) 100w, 240v (c) 75w, 220v (d) 50w, 240v

29. A synchronous motor which is under-excited acts like ____.

(a) a resistor (b) an inductor (c) a capacitor (d) all of these

30. A small lamp used to indicate that a circuit is energized is a ____ lamp.

(a) test (b) pilot (c) electric sign (d) control

31. The power factor of a circuit is approximately 100% if the circuit load consists only of ____.

(a) motors (b) capacitors (c) fluorescent lamps (d) incandescent lamps

32. The length of time between a point in one cycle and the same point in the next cycle of an AC sine wave is the ____.

(a) polarity (b) magnitude (c) period (d) frequency

33. Which of the following does not affect the inductance of a coil?

(a) The diameter of the wire (b) The number of turns
(c) The length of the coil (d) The type of core material

34. The power factor of an AC circuit equals ____.

(a) the cosine of the phase angle (b) the sine of the phase angle
(c) the tangent of the phase angle (d) unity for a reactive circuit

35. A ____ is used to convert DC to AC.

(a) converter (b) inverter (c) power supply (d) multivibrator

36. Reversing the flow in a circuit ____.

(a) changes the voltage (b) destroys the flux
(c) reverses the magnetic polarity (d) does not change the magnetic polarity

37. Disruptive discharges between electrodes of a measuring gap is called ____.

(a) a flashover (b) lightning (c) a sparkover (d) a surge

38. A contact device installed at the outlet for the connection of a single attachment plug is a ____.

(a) junction box (b) rosette (c) reactor (d) receptacle

39. A repulsion motor is equipped with ____.

(a) a commutator **(b) a set of slip rings**
(c) both a commutator and slip ring **(d) neither a commutator nor a slip ring**

40. The rate of rise of current through an inductive coil is maximum ____.

(a) at the start of current flow **(b) near the final maximum value of current**
(c) after one time constant **(d) at 66% of its maximum steady value**

41. The power in a circuit that is not lost in the form of heat is ____.

(a) reactive power (b) true power (c) resistive power (d) none of these

42. Two 1000Ω, one watt resistors are connected in parallel. Their combined resistance and wattage rating is ____.

(a) 500Ω, 1w (b) 500Ω, 2w (c) 1000Ω, 1w (d) 1000Ω, 2w

43. Capacitive reactance opposes changes in ____.

(a) inductance (b) current (c) voltage (d) voltage and current

44. The use of an impressed emf from an external power source used to buck out the corrosion current is called ____ protection.

(a) ionic (b) differential (c) cathodic (d) overcurrent

45. The centrifugal switch of a capacitor start single-phase induction motor is connected in series ____.

(a) with the main windings **(b) with the auxiliary windings**
(c) with the supply terminals **(d) none of these**

46. The result of rust in an electrical connection is ____.

(a) conductance (b) resistance (c) inductance (d) impedance

47. The two things which are the same for the primary and secondary of a transformer are ____.

(a) current and induced voltages **(b) ampere-turns and voltage per turn**
(c) number of turns and power **(d) resistance and leakage reactances**

48. The ____ increases the resistance of wire at high frequencies.

(a) voltage (b) insulation (c) skin effect (d) temperature

49. A lubricant used to make the pulling of wires through conduit easier is ____.

(a) talc (b) resin (c) grease (d) oil

50. When the emfs in the two windings of a transformer are opposite in direction, the polarity of the windings is ____.

(a) subtractive (b) additive (c) either (a) or (b) (d) neither (a) nor (b)

2005

CLOSED BOOK EXAM #8

50 QUESTIONS
TIME LIMIT - 1 HOURS

TIME SPENT ☐ **MINUTES**

SCORE ☐ **%**

MASTER CLOSED BOOK EXAM #8 One Hour Time Limit

1. It is good practice to use ____ in a long horizontal straight run of conduit.

(a) expansion joints (b) flexible joints
(c) universal joints (d) insulation joints

2. The ability of a device to open the maximum short or overload at the device, at a particular point in the electrical system is its ____ capacity.

(a) operating (b) interrupting (c) maximum (d) rated

3. To find the circular-mil area of a stranded conductor, the most accurate method is to ____.

(a) use a micrometer
(b) use a dial-o-meter
(c) square the number of strands
(d) multiply the number of strands by the circular-mil area of one strand

4. All equipment located within ___ feet of the inside wall of a pool shall be grounded.

(a) 15 (b) 10 (c) 8 (d) 5

5. Mica is commonly used for ____.

(a) pole insulators (b) commutator bar separators
(c) panelboards (d) appliance insulation

6. A hook on the end of a fish tape is **not** to ____.

(a) keep it from catching on joints and bends
(b) tie a swab to
(c) tie the wires to be pulled
(d) protect the end of the wire

7. What is the total wattage of the circuit?

(a) 1536 watts (b) 864 watts

(c) 336 watts (d) 192 watts

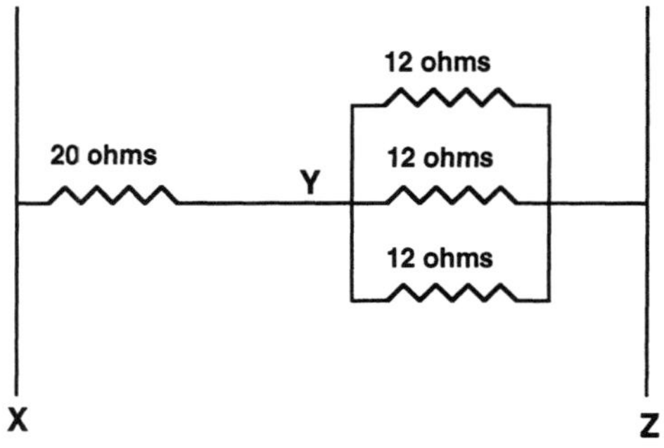

EACH OF THE 12 OHM LOADS IS 2 AMPERES

8. Which of the following statements is **not** an advantage of using an autotransformer?

(a) less expensive
(b) high efficiency
(c) better voltage regulation
(d) safe for stepping down higher voltages

9. ____ is/are the letter(s) used to indicate capacitive reactance in a circuit.

(a) XL (b) Z (c) HZ (d) XC

10. Utilization equipment uses ____.

(a) chemical or mechanical energy **(b) electrical energy only**
(c) energy for electrical purposes **(d) only mechanical energy**

11. A megger is used to measure ____.

(a) reactance (b) specific gravity (c) insulation resistance (d) current

12. A ____ is a device which serves to govern in some predetermined manner the electric power delivered to the apparatus to which it is connected.

(a) switch (b) feeder (c) branch-circuit (d) controller

13. Incandescent light bulbs when operated at a voltage less than the rated voltage, will normally result in ____.

(a) more light
(b) more current
(c) shorter bulb life
(d) longer bulb life

14. The volt-ampere rating in an AC circuit is a way to indicate the ____ power.

(a) true (b) real (c) apparent (d) peak

15. In a three-phase, 4 wire wye system having 277v for lighting, the three-phase motors would be operating on a voltage of ____ volts.

(a) 208 (b) 480 (c) 240 (d) 190

16. Four heater coils with a given voltage will consume the most power when connected ____.

(a) all in series
(b) two in parallel
(c) all in parallel
(d) two parallel pair in series

17. A voltmeter measures ____.

(a) only voltage to ground **(b) voltage difference**
(c) AC voltage only **(d) none of these**

18. In a series circuit with four unknown resistive loads, it is certain that the ____.

(a) voltage drop across each load will be the same
(b) the current flowing through each load will be the same
(c) the wattage consumed by each load will be the same
(d) the total current in the circuit is the sum of all four loads

19. The lead covering on conductors is used ____.

(a) for grounding **(b) for moisture proofing**
(c) for easier soldering **(d) to prevent vibration**

20. The reason for rigid steel conduit to be galvanized or enameled is to ____.

(a) provide better grounding
(b) prevent corrosion
(c) make it easier to thread
(d) make painting easier

21. The N.E.C. covers ____.

I. all X-ray equipment operating at any frequency or voltage for industrial use
II. nonmedical and nondental X-ray
III. radiation safety and performance requirements

(a) I, II and III (b) I and II only (c) I only (d) II only

22. A Plug fuse contains a safety element composed of a piece of metal that has a ____.

(a) low resistance and high melting point
(b) high melting point
(c) low resistance and low melting point
(d) low resistance

23. The definition of automatic is self-acting, operating by its own mechanism when actuated by some impersonal influence such as _____.

(a) a change in current strength **(b) temperature**
(c) mechanical configuration **(d) all of these**

24. DC voltage is reduced with a _____.

(a) resistor (b) capacitor (c) transformer (d) condenser

25. The cord on a heating iron would most likely have an insulation of _____.

(a) nylon (b) rubber (c) asbestos (d) cotton braid

26. If the clips on a cartridge fuse become hot, this is a good indication that _____.

(a) the fuse clips are too loose
(b) the voltage is too high
(c) the fuse clips are too tight
(d) the fuse rating is too high

27. A plastic bushing is considered a(n) _____.

(a) electrical device (b) mounting device (c) fitting (d) insulated device

28. The current will lag the voltage when _____ is present in the circuit.

(a) capacitance (b) inductance (c) reluctance (d) resistance

29. The voltage of a circuit is best defined as _____.

(a) the potential between two conductors
(b) the greatest difference of potential between two conductors
(c) the effective difference of potential between two conductors
(d) the average RMS difference of potential between any two conductors

30. The relationship of a transformer primary winding to the secondary winding is expressed in_____.

(a) wattage (b) volt-amps (c) turns-ratio (d) amps

31. The maximum current a #12 THW copper conductor connected to a 20 amp circuit breaker would ever carry under any condition could be _____ amperes.

(a) 16 (b) 20 (c) 25 (d) 5000 or more

32. The least important reason for using a steel measuring tape around electrical equipment is _____.

(a) magnet effect
(b) danger of entanglement in rotating machinery
(c) shock hazard
(d) short circuit hazard

33. A megawatt is _____ watts.

(a) 100,000 (b) 1000 (c) 10,000 (d) 1,000,000

34. The resistance in a circuit may be increased by _____.

(a) increasing the current flow **(b) increasing the EMF**
(c) a loose connection **(d) decreasing the current flow**

35. D.P.D.T. letters would be used to identify a _____.

(a) circuit breaker (b) type of insulation (c) raceway (d) switch

36. An accessible conductor is _____.

(a) not permanently enclosed by a structure
(b) admitting close approach
(c) not guarded by locked doors, elevation or other methods
(d) being reached without use of a ladder

37. The unit of measure for electrical capacity is the _____.

(a) ohm (b) henry (c) farad (d) watt

38. For better illumination you would use _____.

(a) random spacing of lights **(b) evenly spaced lights, higher ceilings**
(c) even spacing, numerous lights **(d) cluster lights**

39. On a multi-wire, three-wire branch-circuit the maximum unbalanced load on the neutral conductor at anytime would be when _____.

(a) the neutral is disconnected **(b) both circuits are fully loaded**
(c) one hot leg is shut off **(d) both circuits are off**

40. How many 4-way switches are required to switch a light from three places?

(a) one (b) two (c) three (d) four

41. ____ may be connected ahead of the main service switch.

(a) Nothing (b) Lighting fixtures (c) Motors (d) Lightning arresters

42. AC voltages may be increased or decreased by a ____.

(a) rectifier (b) motor (c) transformer (d) dynamo

43. The equipment used to measure and compare peak, average, and root-mean-square values on an AC voltage curve is called a ____.

(a) hydrometer (b) oscilloscope (c) thermal couple (d) mandrel

44. Which of the following is the best conductor of electricity?

(a) iron (b) aluminum (c) tungsten (d) carbon

45. Pure XL in a circuit will cause a ____ degree lag of the ampere curve to the voltage curve in an AC circuit.

(a) 90 (b) 120 (c) 180 (d) 240

46. ____ is/are the letter(s) used to indicate inductive reactance.

(a) XL (b) XC (c) HZ (d) Z

47. With respect to the safety value of insulation on electrical tools, it can be correctly stated that ____.

(a) the insulation should not be used as the only protective measure
(b) the insulation provides very little protection
(c) the insulation provides complete safety to the user
(d) insulation is really not needed

48. A DC voltmeter can be used directly to measure which of the following?

(a) cycles per second (b) power factor (c) power (d) polarity

49. A way of expressing the power factor is ____.

(a) θ (b) R/Z (c) sin θ (d) XL/Z

50. If the line current is 10 amperes with all switches closed, the power consumed in the circuit is ____ watts.

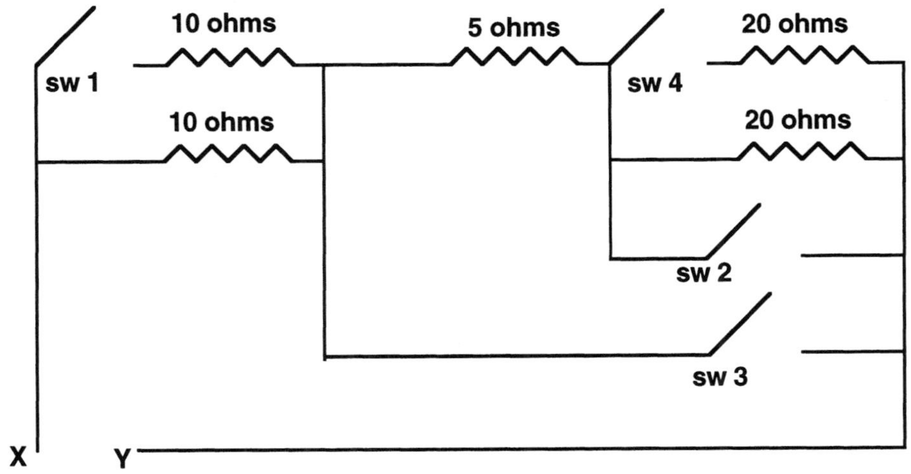

(a) 500 (b) 750 (c) 1000 (d) 2000

2005

CLOSED BOOK EXAM

#9

50 QUESTIONS
TIME LIMIT - 1 HOURS

TIME SPENT ______ **MINUTES**

SCORE ______ **%**

MASTER CLOSED BOOK EXAM #9 One Hour Time Limit

1. When a circuit breaker is selected, the most important factor to consider is the ____ rating.

(a) continuous current (b) interrupting (c) voltage (d) momentary

2. An important feature of a thin film resistor is its ____ temperature coefficient of resistance.

(a) zero (b) low (c) high (d) negative

3. Each component in a motor controller must be approved for which of the following?

(a) The current it must carry
(b) The horsepower that it must control
(c) The voltage to which it will be connected
(d) all of these

4. A good capacitor has a ____ resistance.

(a) very high (b) very low (c) negative (d) negligible

5. Transformers having ratings less than 5 kva are generally ____.

(a) self-cooled (b) water cooled (c) oil cooled (d) natural air cooled

6. The winding of the transformer with the greater number of turns will be _____.

(a) the high-voltage winding
(b) the low voltage winding
(c) the high power winding
(d) either the high or low voltage winding

7. The value of the voltage that is indicated on an AC voltmeter is called the ____.

(a) useful value (b) average value (c) maximum value (d) effective value

8. A ground should have a ____ resistance.

(a) low (b) moderate (c) infinite (d) high

9. If two similar conductors are connected in parallel, their total resistance is ____.

(a) doubled
(b) cut by one third
(c) half the resistance of one wire
(d) the resistance of one wire

10. A coil connected to a DC source will exhibit ____.

(a) resistance and inductance (b) only DC resistance (c) only inductance (d) any of these

11. A luminous discharge due to ionization of the air surrounding a conductor caused by a voltage gradient exceeding a certain critical value is called ____.

(a) surge (b) flashover (c) skin effect (d) corona

12. An auxiliary conductor used in connection with remote measuring devices or for operating apparatus at a distant point is a ____ wire.

(a) pilot (b) jumper (c) tie (d) connector

13. A sensor is used for ____.

(a) converting an electrical input to mechanical output
(b) increasing the band width of tuned circuits
(c) converting a non-electrical input to electrical output
(d) reducing the effects of noise and other disturbances on the measured quantity

14. An AC electromagnet will ____.

(a) repel other magnetized objects
(b) attract pure, unmagnetized iron
(c) attract only other magnetized objects
(d) either attract or repel permanent magnets, depending on polarity

15. A low voltage source such as 12 volts _____.

(a) is not dangerous at all (b) can be dangerous under certain conditions
(c) is always dangerous (d) is dangerous if it is AC but not if DC

16. The torque developed by a three-phase motor depends on the following three factors ____.

(a) speed, frequency, and number of poles (b) rotor emf, rotor current, and rotor PF
(c) voltage, current, and stator impedance (d) magnetic flux, eddy currents, and emf

17. An AC line voltage of 120 volts has a peak value of approximately ____ volts.

(a) 84 (b) 105 (c) 170 (d) 339

18. The capacitor used in power factor correction is rated in ____.

(a) kw (b) kva (c) kvar (d) emf

19. When a motor is to be controlled from two different locations, the start buttons are connected in ____ while the stop buttons are connected in ____.

(a) series, parallel (b) series, series (c) parallel, parallel (d) parallel, series

20. A relay which operates and resets with no intentional time delay is called a _____ relay.

(a) inverse-time (b) delay-off (c) electromechanical (d) instantaneous-trip

21. A good fuse should have ____ resistance.

(a) approximately no (b) a very high (c) neither a nor b (d) either a or b

22. A ____ lamp is basically a mercury lamp with a certain innovation of its arc tube.

(a) metal halide (b) neon (c) sodium (d) incandescent

23. Which of the following is a variable speed motor?

(a) Series (b) Shunt (c) Differentially compound (d) Commulatively compound

24. A transformer works on ____.

(a) DC only (b) AC only (c) AC as well as DC (d) pulsating DC

25. Another name for full voltage motor starting is ____.

(a) full load starting (b) high voltage starting (c) direct on line (d) maximum starting

26. A ____ transforms heat energy to electric energy.

(a) generator (b) thermocouple (c) battery (d) transformer

27. The current carrying capacity of the fuse material depends on ____.

(a) length (b) cross-sectional area (c) material (d) all of these

28. A 220 volt, 10 hp, single-phase induction motor operates at an efficiency of 86% at a power factor of 90%. The current would be ____ amps.

(a) 43.81 (b) 35.7 (c) 42.3 (d) 46.1

29. Which one of the following parameters cannot be changed by a transformer?

(a) voltage (b) current (c) power (d) impedance

30. The opposition to the creation of magnetic lines of force is ____.

(a) susceptance (b) conductance (c) elastance (d) reluctance

31. The power factor of an inductive circuit can be corrected by adding ____.

(a) a capacitor (b) more inductors (c) filters (d) diodes

32. A voltmeter should have ____.

(a) a high internal resistance (b) a low internal resistance
(c) a medium internal resistance (d) none of these

33. A unity power factor is indicative that the circuit is ____.

(a) reactive (b) purely inductive (c) purely capacitive (d) resistive

34. What does "PH" mean on the nameplate of a motor?

(a) poly-hertz (b) per horsepower (c) number of phases (d) phase heat

35. In measuring high resistance values such as insulators, normally you would use a/an ____.

(a) ohmmeter (b) megger (c) Wheatstone bridge (d) rotometer

36. The starting winding of a single-phase motor is placed in the ____.

(a) field (b) stator (c) rotor (d) armature

37. When an alternating voltage reverses in polarity, the current it produces ____.

(a) reverses in direction (b) has a steady DC value
(c) has a phase angle of 180° (d) none of these

38. The type of overload that contains a solder pot is called _____.

(a) melting alloy (b) metallic (c) bimetallic (d) magnetic

39. If the three-phase load is balanced, the minimum number of wattmeters required to measure the power is ____.

(a) one (b) two (c) three (d) four

40. The RMS value of an alternating wave is the same as the ____ value.

(a) absolute (b) average (c) instantaneous (d) effective

41. A method for stopping a three-phase motor quickly by momentarily connecting the motor for reverse rotation is called ____.

(a) plugging (b) inching (c) jogging (d) latching

42. A reverse-power relay is used to ____.

(a) change AC power to DC power
(b) protect the generator from power reversal
(c) reverse the rotation of the motor
(d) protect the motor from power surges

43. The magnetizing current for the field of an alternator is usually supplied by a ____.

(a) pulse generator (b) battery (c) DC magnet (d) DC generator

44. An autotransformer is used in starting a large induction motor in order to ____.

(a) reduce the line voltage at starting
(b) step-up the line voltage at starting
(c) to apply dynamic braking
(d) to accelerate the clutch action

45. A centrifugal switch used in split-phase motors will cut-off the starting winding after the motor developed approximately ____ of its rated speed.

(a) 70% (b) 75% (c) 80% (d) 85%

46. A resistance wire wrapped around an insulating core is ____.

(a) film-type (b) fusible type (c) wire wound (d) carbon composition

47. During the open circuit test on transformers, the ____ is opened.

(a) high (b) low (c) both a & b (d) either a or b

48. A portion of a lightning protection system extending into earth is a ____.

(a) ground terminal (b) counterpoise (c) air terminal (d) surge arrester

49. Transformers used to step voltages down for general use are classified as ____ systems.

(a) emergency (b) direct (c) stand-by (d) separately derived

50. Where knobs are used, conductors shall be securely tied thereto by ____ wires having insulation equivalent to that of the conductor.

(a) copper (b) guy (c) aluminum (d) tie

2005

CLOSED BOOK EXAM #10

50 QUESTIONS
TIME LIMIT - 1 HOURS

TIME SPENT ☐ **MINUTES**

SCORE ☐ **%**

MASTER CLOSED BOOK EXAM #10 **One Hour Time Limit**

1. Solenoids are made of what type of magnets?

(a) reverse (b) permanent (c) electro (d) copper

2. The switches to be closed in order to obtain a combined resistance of 5 ohms are ___ switches.

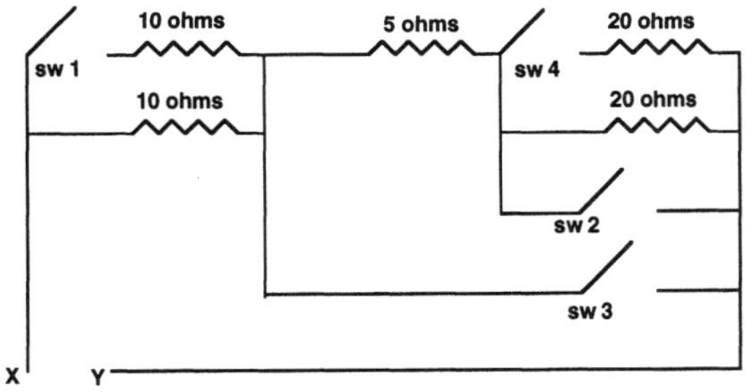

(a) 1 and 3 **(b) 2 and 3**

(c) 1 and 2 **(d) 1 and 4**

3. To measure the RPM speed of a motor you would use a ____.

(a) hydrometer (b) bolometer (c) tachometer (d) odometer

4. Which of the following is **not** a factor in calculating the feeder conductor size?

(a) ambient temperature **(b) branch-circuit protection**
(c) voltage drop **(d) demand factors**

5. ____ is a synthetic non-flammable insulating liquid, which when decomposed by electrical arcs, involve non-flammable gases.

(a) Prynol (b) Askarel (c) Electrolyte (d) Hermetic

6. Voltage drop in a conductor is ____.

(a) the conductor resistance times the voltage
(b) a function of insulation
(c) part of the load voltage
(d) a percentage of the applied voltage

7. The ratio of the maximum demand of the system to the total connected load of the system is called the ____ of the system.

(a) connected load (b) nameplate (c) demand factor (d) turns-ratio

8. Piezoelectric is caused by crystals or binding ____.

(a) heat (b) pressure (c) static (d) chemical

9. Since fuses are rated by amperes and voltage, a fuse will work on ____.

I. AC II. DC III. any voltage level

(a) I only (b) II only (c) III only (d) either I or II

10. When three equal resistors are connected in parallel the total resistance is ____.

(a) the sum of the three resistors
(b) one-half of the sum of the three resistors
(c) less than any one resistor
(d) greater than any two equal resistors

11. Light fixtures hung by chains should be wired so that the ____.

(a) chain is not grounded
(b) wires support the light
(c) wires do not support the light
(d) light is insulated from the chain

12. ____ boxes may be weatherproof.

I. Rainproof II. Raintight III. Watertight

(a) I only (b) II only (c) III only (d) I, II and III

13. Conductor overheating can be caused by ____.

I. overloading
II. high ambient temperature
III. grouping of more than 3 current-carrying conductors without derating

(a) I only (b) II only (c) II and III only (d) I, II or III

14. A switch or breaker should disconnect the grounded conductors of a circuit ____.

(a) only with an isolating switch
(b) before disconnecting the hot conductors
(c) simultaneously as it disconnects the ungrounded conductors
(d) the grounded conductor can never be switched

15. The definition of ambient temperature is ____.

(a) the temperature of the conductor
(b) the insulation rating of the conductor
(c) the temperature of the area surrounding the conductor
(d) the maximum heat the insulation can be used within

16. For voltage and current to be in phase ____.

(a) voltage and current appear at their zero and peak values at the same time
(b) the circuit impedance has only resistance
(c) neither (a) nor (b)
(d) both (a) and (b)

17. Which of the following instruments would you use to indicate the phase relationship between the voltage and the current of an AC circuit?

(a) KWH meter (b) power factor meter (c) manometer (d) growler

18. A ____ conductor is one having one or more layers of non-conducting materials that are not recognized as electrical insulation.

(a) wrapped (b) bare (c) covered (d) insulated

19. Electrical equipment can be defined as ____.

I. fittings II. appliances III. devices IV. fixtures

(a) I only (b) I and IV only (c) I, III and IV (d) I, II, III and IV

20. The definition of a dry location ____.

(a) not normally subjected to dampness
(b) not normally subjected to wetness
(c) may be temporarily subjected to wetness
(d) all of the above

21. ____ duty is a type of service where both the load and the time intervals may have wide variations.

(a) Continuous (b) Periodic (c) Intermittent (d) Varying

22. An autotransformer is generally used rather than an isolation transformer ____.

(a) when cost is a factor
(b) where the ratio of transformation is low
(c) when you have several branch-circuits
(d) when safety is a factor

23. The symbol for a wye connection is ____.

(a) Y (b) Ω (c) ø (d) Δ

24. A ____ overcurrent device is not designated to interrupt short-circuit current.

(a) inverse breaker (b) dual-element fuse (c) type S fuse (d) thermal cutout

25. If a test lamp lights when placed in series with a condenser and a suitable source of DC, it is a good indication that a condenser is _____ if the light burns continuously.

(a) fully charged **(b) short circuited**
(c) open circuit **(d) fully discharged**

26. Impedance is measured in ____.

(a) farads (b) henrys (c) ohms (d) coloumbs

27. The standard unit of electrical pressure is the ____.

(a) ampere (b) watt (c) volt (d) watt-hour

28. The common fuse depends on the principle that the ____.

(a) current flow developes heat **(b) overvoltage will expand the link**
(c) increase of resistance will occur **(d) voltage developes heat**

29. The scope of the National Electrical Code includes ____.

(a) conductors and equipment wherever and whenever electricity is employed
(b) all buildings or premises where electrical power is employed
(c) conductors and equipment on certain public or private premises
(d) all of these

30. Electro-magnetic devices usually have a ____ core.

(a) aluminum (b) soft iron (c) hard iron (d) hard steel

31. When a current leaves its intended path and returns to the source, bypassing the load, the circuit is ____.

(a) broken (b) open (c) shorted (d) incomplete

32. The most heat is created when current flows through which of the following?

(a) a 10 ohm inductance coil
(b) a 10 ohm resistor
(c) a 10 ohm condenser
(d) the heat would be equal in all of these

33. The greater number of free electrons, the better the ____ of a metal.

(a) voltage drop (b) resistance (c) conductivity (d) insulation value

34. When one material gives up electrons readily and another attracts them, this is the basis for ____.

(a) piezoelectric (b) thermocouple (c) electrochem (d) photovoltaic

35. The voltage will lead the current when ____.

(a) capacitive reactance exceeds the inductive reactance in the circuit
(b) resistance exceeds reactance in the circuit
(c) reactance exceeds the resistance in the circuit
(d) inductive reactance exceeds the capacitive reactance in the circuit

36. The vector sum of the phase currents is equal to what in a balanced, resistive three-phase system?

(a) phase current x power factor **(b) zero**
(c) 1.732 x phase current **(d) three x phase current**

37. The heating of two different metals will cause ____.

(a) corrosion (b) electron flow (c) galvanic action (d) fusion

38. If you are using two wattmeters to measure the power, you are measuring power in a ____ system.

(a) single-phase (b) three-phase (c) double (d) high-voltage

39. When referring to a "8-32" machine bolt, the "32" refers to the ____.

(a) threads per inch (b) length of the bolt (c) diameter (d) strength

40. On a DC ammeter, to increase the range you would use a ____.

(a) capacitor (b) inductor (c) shunt (d) condenser

41. A motor works in the principles of ____.

(a) magnetism (b) mechanical force (c) residual force (d) chemical action

42. Connecting batteries in series will give greater ____.

(a) amp hours (b) amperage (c) voltage (d) all of these

43. In a completed seal, the minimum thickness of the sealing compound shall not be less than the trade size of the conduit, and in no case less than ____ inch.

(a) 1/4 (b) 3/8 (c) 1/2 (d) 5/8

44. Root-mean-square is also called the ____ voltage.

(a) average (b) peak (c) effective (d) maximum

45. The current through a DC arc lamp can be controlled by a ____.

(a) transformer (b) rheostat (c) rectifier (d) scanner drum

46. The current used for charging storage batteries is ____.

(a) direct (b) positive (c) alternating (d) negative

47. When threading rigid conduit the standard die should provide ____ taper per foot.

(a) 1/4" (b) 1/2" (c) 3/4" (d) 1"

48. Lubrication would never be applied to a ____.

(a) bearing (b) knife switch (c) controller drum (d) commutator

49. When the power factor in a given circuit is unity, the reactive power is ____.

(a) at maximum (b) 1.1414 (c) zero (d) a negative quantity

50. Materials containing numerous free electrons are ____.

(a) good insulators (b) ferrous alloys (c) good conductors (d) carbons

2005

CLOSED BOOK EXAM #11

50 QUESTIONS
TIME LIMIT - 1 HOURS

TIME SPENT [ ] MINUTES

SCORE [ ] %

MASTER CLOSED BOOK EXAM #11 **One Hour Time Limit**

1. The current will lead the voltage when ____.

(a) inductive reactance exceeds the capacitive reactance in the circuit
(b) reactance exceeds the resistance in the circuit
(c) resistance exceeds reactance in the circuit
(d) capacitive reactance exceeds the inductive reactance in the circuit

2. The difference between a thread on a bolt and a thread on a conduit is ____.

(a) the conduit thread is deeper
(b) the conduit thread is tapered
(c) no lubrication required in threading conduit
(d) the conduit thread always has the same pitch for any diameter

3. Open conductors run individually as service drops shall be ____.

I. insulated II. bare III. covered

(a) I only (b) II only (c) III only (d) either I or III

4. A ____ is a function that is similar to a rectifier.

(a) commutator (b) transformer (c) contactor (d) inverter

5. The term pothead as used in the electrical trade means ____.

(a) current-limiting fuse **(b) cable terminal**
(c) protective device **(d) insulator on a line pole**

6. If an AC sine wave reaches a peak voltage of 100, what is the effective root-mean square voltage?

(a) 57.7 volts (b) 141.4 volts (c) 86.6 volts (d) 70.7 volts

7. What is the total wattage of this circuit?

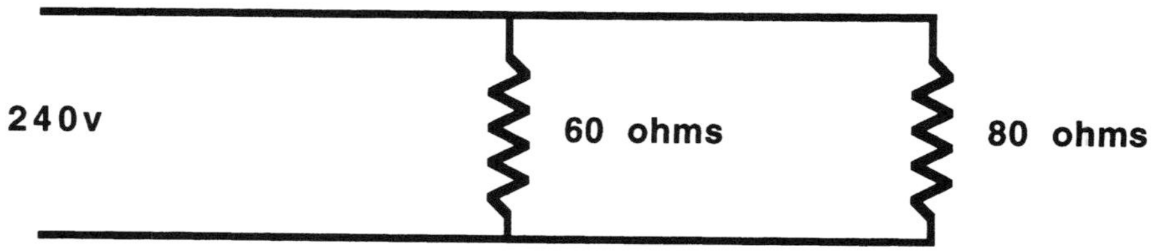

(a) 3.5 (b) 420 (c) 16,800 (d) 1680

8. When measuring to determine the size of a stranded conductor, you would place the wire guage over ____.

(a) one strand of the conductor **(b) the insulation**
(c) all of the strands **(d) the outer covering**

9. The resistance of the filament in a light bulb is ____.

(a) usually the same at all times **(b) highest when bulb is off**
(c) lowest when bulb is on **(d) highest when bulb is on**

10. When one electrical circuit controls another electrical circuit through a relay the circuit is a ____ circuit.

(a) remote-control (b) control (c) primary (d) secondary

11. Two one ohm resistors in parallel, the total R is ____ ohm(s).

(a) 1 (b) 2 (c) 1/2 (d) cannot be calculated

12. The advantage(s) of a 4-wire, three-phase source over a 3-wire, three-phase source is/are ____.

I. two voltage levels II. a grounded neutral III. less copper required

(a) I only (b) II only (c) III only (d) I and II

13. An ammeter is disconnected from a current transformer (CT), the leads should be ____.

(a) taped (b) shorted (c) re-routed (d) insulated

14. When working on live 600 volt equipment where rubber gloves might be damaged, an electrician should ____.

(a) wear leather gloves over rubber ones
(b) work without gloves
(c) reinforce the fingers with tape
(d) carry a spare pair

15. A fuse puller is used to replace ____ fuses.

(a) cartridge (b) plug (c) link (d) current-limiting

16. Aluminum and copper-clad aluminum of the same circular-mil and insulation have ____.

(a) the same physical characteristics **(b) the same termination methods**
(c) the same ampacity **(d) different ampacities**

17. The primary purpose for grounding a raceway is to prevent the raceway from becoming ____.

(a) accidentally energized at a higher potential than ground
(b) a source of induction
(c) magnetized
(d) a path for eddy currents

18. When operated on a voltage 10% higher than nameplate rating, an appliance will:

I. have a shorter life II. draw a higher current III. use more power

(a) I only (b) II only (c) III only (d) I, II and III

19. The liquid in a battery is called the ____.

(a) askarel (b) eddy current (c) hermetic (d) electrolyte

20. ____ is current which flows through a circuit in the same direction at all times and with almost constant strength.

(a) AC (b) Wattage (c) DC (d) Ampacity

21. An Erickson coupling is used to ____.

(a) join sections of EMT together
(b) connect EMT to flexible conduit
(c) connect two sections of rigid conduit when one section cannot be turned
(d) substitute for all-thread

22. To fasten a box to a terra cotta wall you would use ____.

(a) lag bolts (b) expansion bolts (c) wooden plugs (d) rawl plugs

23. If voltage rotation of a three-phase motor is A-B-C, the current rotation is ____.

(a) C-B-A (b) A-C-B (c) B-C-A (d) B-A-C

24. The disadvantage of an autotransformer is ____.

(a) the high cost
(b) the poor efficiency due to a single winding
(c) the size
(d) the lack of isolation between the primary and secondary

25. The power factor can be determined by the use of which combination of the following meters?

I. ammeter II. wattmeter III. watt-hour meter IV. ohmmeter
V. voltmeter VI. phase rotation meter

(a) I, V and VI (b) I, IV and VI (c) I, II and V (d) none of these

26. A unit of an electrical system which is intended to carry but not utilize electrical energy is a(an) ____.

(a) device (b) motor (c) light bulb (d) appliance

27. When relay contacts make and break frequently, a condenser is used across the relay contacts to ____.

(a) energize the relay quicker
(b) delay the coil energizing
(c) reduce pitting of the contacts
(d) increase the efficiency

28. The main reason for using oil in an oil circuit breaker is to ____.

(a) lubricate the points
(b) quench the arc
(c) increase the capacity of current
(d) decrease the resistance

29. The resistance of a copper conductor is ____.

(a) inversely proportional to its cross-sectional area
(b) inversely proportional to its length
(c) directly proportional to its diameter
(d) directly proportional to the square of its diameter

30. To measure AC cycles per second, you would use a ____.

(a) hydrometer (b) manometer (c) frequency meter (d) power factor meter

31. Voltaic current will lead the voltage when ____.

(a) chemical (b) solar (c) both of these (d) none of these

32. Inductance is measured in ____.

(a) farads (b) henrys (c) coulombs (d) amperes

33. A motor with a wide speed range is a(n) ____.

(a) DC motor (b) AC motor (c) synchronous motor (d) induction motor

34. For discharge lighting lamps to operate properly they would require ____.

(a) some sort of ballast (b) a starter
(c) resistive starter (d) a cool ambient

35.Comparing incandescent lighting with fluorescent lighting at the same level of illumination, the cost of energy for fluorescent would be ____.

(a) greater (b) less (c) the same (d) 1.05 increase

36. Motor horsepower ratings are based on an observable safe temperature rise above ambient temperature. The ambient temperature is taken as ____ degrees C.

(a) 40 (b) 45 (c) 50 (d) 60

37. In a three-phase circuit, the legs of the phase are ____ electrical degrees apart.

(a) 90 (b) 180 (c) 120 (d) 240

38. A generator works on the basis of ____.

(a) a conductor moving inside a magnetic field (b) ozone principle
(c) a magnetic field moving around a conductor (d) none of these

39. In an area that requires explosion-proof wiring, raceways entering a box in this area which contains equipment that may produce sparks, shall be provided with ____.

(a) an approved sealing compound (b) insulated bushings
(c) two grounding conductors (d) triple lock nuts

40. In a 60 cycle system, what length of time does it take to go 90 degrees?

(a) 1/3 second (b) 1/90 second (c) 1/60 second (d) 1/240 second

41. The basic unit of electrical work is the _____.

(a) volt-amp (b) watt (c) watt-hour (d) kva

42. A good insulator for very high temperature is _____.

(a) mica (b) rubber (c) plastic (d) bakelite

43. In hazardous locations which is/are factors that contribute to the need to classify the location as hazardous?

I. flammable liquids II. certain dust particles III. proximity to local fire station

(a) I only (b) II only (c) I or II only (d) I, II or III

44. The main reason for laminating the core of a power transformer is to keep the _____.

(a) copper losses at a minimum (b) turns-ratio balanced
(c) eddy current loss at a minimum (d) friction losses low

45. To open a single-throw switch, the switch should be mounted so that the switch blade must move _____.

(a) downward (b) upward (c) twice (d) slowly

46. The total opposition to current flow in an AC circuit is expressed in ohms as _____.

(a) impedance (b) conductance (c) reluctance (d) resistance

47. Which of the following motors does **not** have a commutator?

(a) squirrel cage (b) series (c) shunt (d) synchronous

48. Resistance can be found directly from wattage and voltage measurements by the equation _____.

(a) $R = V^2/W$ (b) $R = V/W$ (c) $R = W/V$ (d) $R = W^2/V$

49. A coil has 100 turns. A current of 10 amps is passed through the coil. The coil developes _____.

(a) 1000 watts (b) 1 kva (c) mutual inductance (d) 1000 amp turns

50. What is the voltage drop across the 20 ohm series load?

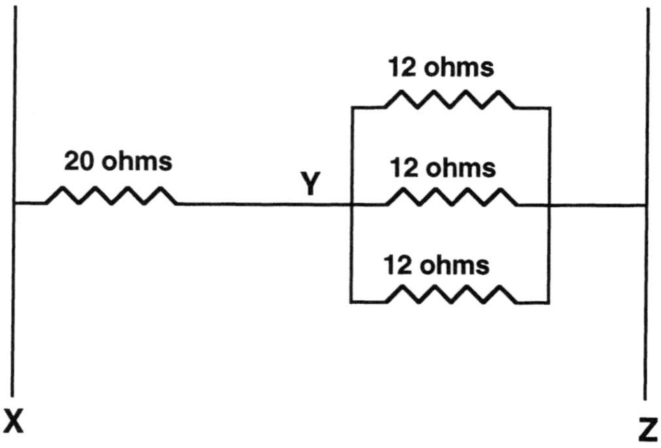

EACH OF THE 12 OHM LOADS IS 2 AMPERES

(a) 144 volts (b) 24 volts (c) 120 volts (d) 336 volts

2005

CLOSED BOOK EXAM

#12

50 QUESTIONS
TIME LIMIT - 1 HOURS

TIME SPENT ☐ **MINUTES**

SCORE ☐ **%**

MASTER CLOSED BOOK EXAM #12 — One Hour Time Limit

1. If the phase voltages are 120 degrees apart, the line voltages will be ____ degrees apart.

(a) 30 (b) 60 (c) 90 (d) 120

2. Electrical current is measured in terms of ____.

(a) electrical pressure **(b) electrons passing a point per second**
(c) watts **(d) ohms**

3. A ____ is used for testing specific gravity.

(a) galvanometer (b) mandrel (c) calorimeter (d) hydrometer

4. In the course of normal operation the instrument which will be **least** effective in indicating that a generator may overheat because it is overloaded is ____.

(a) a stator thermocouple **(b) a wattmeter**
(c) a voltmeter **(d) an ammeter**

5. An outlet box should be fastened to a solid concrete wall by means of ____.

(a) toggle bolts (b) wood plugs (c) lag bolts (d) expansion bolts

6. What would be the advantage of 240 volts rather than 120 volts on a load with the same wattage?

(a) less power used **(b) less voltage drop**
(c) both (a) and (b) **(d) neither (a) nor (b)**

7. The cross-sectional area of the bus bar is ____ square inch.

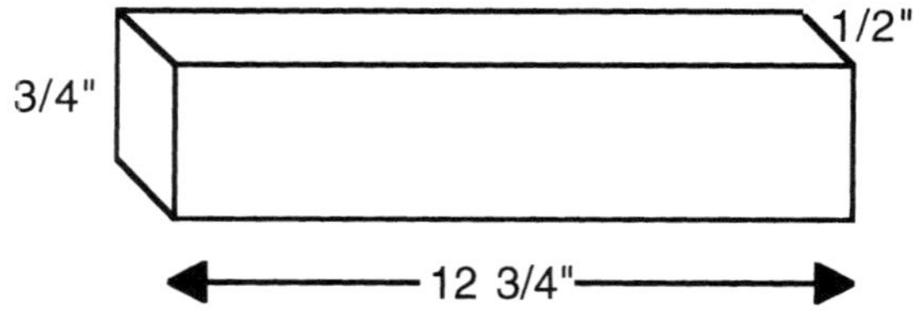

(a) .375 (b) 4.78125 (c) 14 (d) none of these

8. AC voltage of the system is **not** the ____ voltage.

(a) EMF (b) effective (c) average (d) RMS

9. A multimeter is a combination of ____.

(a) ammeter, ohmmeter and wattmeter
(b) voltmeter, ohmmeter and ammeter
(c) voltmeter, ammeter and megger
(d) voltmeter, wattmeter and ammeter

10. Two 120 volt light bulbs connected in series across 240 volts will ____.

(a) burn at full brightness **(b) burn at half-brightness**
(c) burn out quickly **(d) flicker with the cycle**

11. The N.E.C. covers ____.

I. audio signal generation II. speech-input systems III. electronic organs

(a) I, II and III (b) I and II only (c) I only (d) III only

12. The conductor used to connect the grounded circuit of a wiring system to a grounding electrode is the ____.

(a) grounded conductor **(b) bonding jumper**
(c) main bonding jumper **(d) grounding conductor**

13. Galvanized rigid conduit is made of ____.

(a) iron (b) zinc (c) lead (d) nickel

14. A soldering iron should not be heated to excess because it will usually ____.

(a) ruin the tin on the surface of the tip **(b) burn the wooden handle**
(c) short-circuit the iron **(d) anneal the copper tip**

15. A pendant fixture is a ____.

(a) hanging fixture **(b) recessed fixture**
(c) bracket fixture **(d) none of these**

16. ____ means so constructed or protected that exposure to the weather will not interfere with successful operation.

(a) Weatherproof (b) Weather-tight (c) Weather-resistant (d) Weather-sealed

17. To reverse the rotation of a three-phase motor you would ____.

(a) turn it around
(b) reverse two of the four leads
(c) reverse any two of the three leads
(d) reverse all the leads

18. The disadvantage of AC compared to DC would be ____.

(a) smaller conductors to transmit electrical energy
(b) reduced cost of transmission with high voltage transformers
(c) less copper required with AC at higher voltages
(d) cannot be used in electro-plating

19. XL, XC, and R each in its own branch-circuit will flow in ____ directions.

(a) one (b) two (c) three (d) none of these

20. In a highly inductive AC circuit, what device is used to correct the power factor towards unity?

(a) resistor (b) inductor (c) capacitor (d) rectifier

21. In other than residential calculations, an ordinary outlet shall be calculated at ____.

(a) 660va (b) 746w (c) 2 amps (d) 180va

22. New brushes for a generator should be fitted to the commutator by using a ____.

(a) bastard file
(b) strip of emery cloth
(c) polishing stone
(d) strip of sandpaper

23. Wound rotor and squirrel cage are two types of ____motors.

(a) synchronous (b) single-phase (c) induction (d) three-phase

24. A galvanometer is used to indicate ____.

(a) voltage (b) current (c) resistance (d) wattage

25. The sum of the voltage drop around a circuit is equal to the source voltage is ____.

(a) Kirchhoff's Law (b) Ohm's Law (c) Watt's law (d) Nevin's Theory

26. Ohm's Law states ____.

(a) E is equal to I/R **(b) E is equal to R/I**
(c) E is equal to I times R **(d) none of these**

27. A clamp-on ammeter will measure ____.

(a) voltage when clamped on a single conductor
(b) current when clamped on a multiconductor cable
(c) accurately only when parallel to cable
(d) accurately only when clamped perpendicular to a conductor

28. The purpose of locknuts in making electrical connections on studs is to ____.

(a) prevent the connection from loosening under vibration
(b) connect multiple conductors on the stud
(c) make the connection tamperproof
(d) avoid having to torque the studs

29. What winding of a current transformer will carry more current?

(a) primary (b) secondary (c) interwinding (d) tertiary

30. The words "thermally protected" appearing on the nameplate of a motor indicates that the motor is provided with a ____.

(a) switch (b) fuse (c) breaker (d) heat sensing element

31. What type of meter is shown in the diagram?

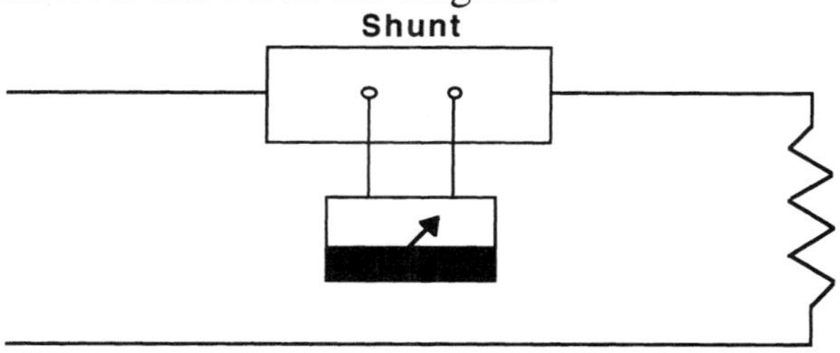

(a) wattmeter (b) ammeter (c) ohmmeter (d) voltmeter

32. True power is always voltage times current ____.

(a) in an AC circuit
(b) in a DC circuit
(c) where frequency is constant
(d) regardless of whether capacitive reactance is in the circuit or not

33. Which of the following would have the least amount of resistance?

(a) semi-conductor (b) insulator (c) resistor (d) conductor

34. Grounding the metallic cover of flexible metal conduit and armored cable, is for protection against ____.

(a) shock or injury
(b) lightning
(c) open field shorts
(d) change of frequency

35. Which of the following breaks down rubber insulation?

(a) oil (b) acid (c) water (d) none of these

36. The opposite of resistance would be ____.

(a) resonance (b) reluctance (c) reactance (d) conductance

37. Conductors, such as non-metallic cable, when run in the space between studs are defined as ____.

(a) inaccessible (b) concealed (c) enclosed (d) buried

38. A type of transformer that has only a single winding is called a ____ transformer.

(a) tertiary (b) auto (c) isolation (d) saturated

39. The main reason that water should not be used to extinguish fires involving electrical equipment is that water ___.

(a) may transmit shock to the user
(b) is ineffective in extinguishing electrical fires
(c) may short out the equipment
(d) will ruin the electrical equipment

40. Which of the following grounding electrodes is the only one that shall be supplemented by an additional electrode?

(a) metal underground water pipe (b) ground ring (c) building steel (d) concrete-encased

41. In a multiple motor circuit there is feeder protection, branch-circuit protection and motor protection. If the feeder protection trips, the fault would be expected to be in ____.

(a) one of the motors **(b) one of the branch-circuits**
(c) the feeder **(d) one of the starters**

42. A branch-circuit that supplies a number of outlets for lighting and appliances is known as a ____ branch-circuit.

(a) general purpose (b) multi-purpose (c) utility (d) none of these

43. The advantage of AC over DC includes which of the following?

(a) better speed control **(b) lower resistance at high currents**
(c) ease of voltage variation **(d) impedance is greater**

44. When three light bulbs are wired in a single fixture, they are connected in ____.

(a) parallel (b) series (c) series-parallel (d) none of these

45. A voltage of 2000 microvolts is the same as ____ volts.

(a) 0.002 (b) 0.200 (c) 0.020 (d) 2,000

46. Class ___ locations are those that are hazardous because of the presence of easily ignitible fibers or flyings.

(a) I (b) II (c) III (d) none of these

47. ____ conductors shall be used for wiring on fixture chains.

(a) Solid (b) Bare (c) Covered (d) Stranded

48. The proper way to mount an instrument meter is to ____.

(a) use a template **(b) use the meter to mark your holes**
(c) drill from the back **(d) drill from the front**

49. Three conductors (phase A, B and neutral) run from a busway to a load center should **not** be installed in three separate raceways because ____.

(a) you need room for future fill **(b) to avoid using a gutter**
(c) of inductance **(d) its easier to pull the conductors**

50. If 3 amperes flow through the 5 ohm resistor with all switches open, the voltage between the terminals X and Y is ____ volts.

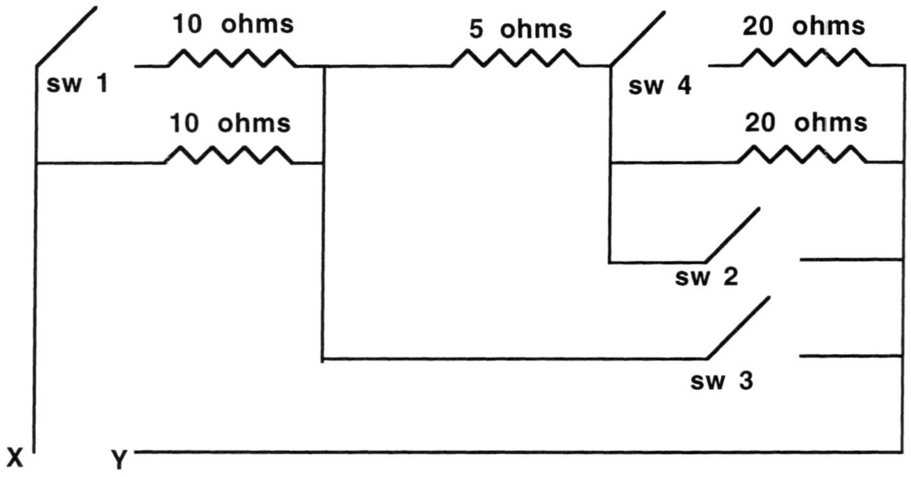

(a) 15 (b) 60 (c) 90 (d) 105

2005

CLOSED BOOK EXAM #13

50 QUESTIONS
TIME LIMIT - 1 HOURS

TIME SPENT ______ **MINUTES**

SCORE ______ **%**

MASTER CLOSED BOOK EXAM #13 **One Hour Time Limit**

1. The tool that is used to align vitrified tile conduit in multiple ducts is a ____.

(a) mandrel (b) manometer (c) growler (d) hickey

2. If frequency is constant, the inductive reactance of a circuit will ____.

(a) remain constant regardless of voltage or current change
(b) vary with voltage
(c) vary directly with current
(d) not effect the impedance

3. An isolating switch is one that is ____.

(a) intended for cutting off an electrical circuit from its source of power
(b) required to have a padlock
(c) primarily used with an isolation transformer
(d) used only for heavy motor overloads

4. The definition of a qualified person is ____.

(a) persons designated by the management of a building to supervise or work on equipment
(b) persons with a degree or license qualifying them to install, maintain, or operate a specific type of equipment or system
(c) a Master electrician
(d) persons familiar with the construction and operation of the equipment and the hazards

5. ____ is a system or circuit conductor that is intentionally grounded.

(a) Grounding conductor **(b) Grounded conductor**
(c) Neutral conductor **(d) Grounding electrode conductor**

6. Solders used for the connection of electrical conductors, are alloys of ____.

(a) tin and lead (b) copper and tin (c) zinc and lead (d) zinc and tin

7. If the maximum current on a circuit is 70 amperes, the ammeter will read  ?

(a) 70 amps (b) 60.4 amperes (c) 49.49 amperes (d) 40.62 amperes

8. Two 500 watt lamps connected in series across a 120 volt source draws 3 amps, the total wattage consumed is ____.

(a) 1000 watts (b) 500 watts (c) 360 watts (d) 200 watts

9. Raceways on the outside of buildings should be ____.

(a) weatherproof and covered
(b) watertight, arranged to drain
(c) raintight and arranged to drain
(d) rainproof and guarded

10. In a three-phase system with 4-wire, 208/120 volt, there exists ____.

(a) 3 ø 208 volt, 3 ø 120 volt, and 1 ø 120 volt
(b) 3 ø 208 volt, 1 ø 120 volt, and 1 ø 208 volt
(c) 3 ø 208 volt, 3 ø 120 volt, and 1 ø 208 volt
(d) 1 ø 208 volt, 3 ø 208 volt, and 3 ø 120 volt

11. Which of the following is best to fight electrical fires?

(a) soda-acid fire extinguisher
(b) fine spray of water
(c) foam fire extinguisher
(d) CO_2 fire extinguisher

12. A substance that would be good as an electrical insulation is which of the following?

(a) carbon (b) oil (c) lead (d) iron

13. The letters CATV refer to ____.

(a) communications (b) TV (c) antenna systems (d) video

14. If the allowable current in a copper bus bar is 1000 amperes p.s.i. of cross-section, the width of a standard 1/4" bus bar designed to carry 1500 amperes would be ____.

(a) 2" (b) 4" (c) 6" (d) 8"

15. The NEC requires that conduit must be continuous from outlet to outlet, must be mechanically and electrically connected to all fittings, and must be suitably grounded. The reason for having the conduit electrically continuous and grounded is to ____.

(a) shield the wires inside the conduit from external magnetic fields.
(b) provide a metallic return conductor.
(c) make it easy to test wiring connections.
(d) prevent electrical shock which might otherwise result from contact with the conduit.

16. High AC voltages are usually measured using a ____.

(a) potential transformer and voltmeter
(b) current transformer and a voltmeter
(c) galvanometer in parallel
(d) manometer in series with a voltmeter

17. What is the wattage of the three 12 ohm loads?

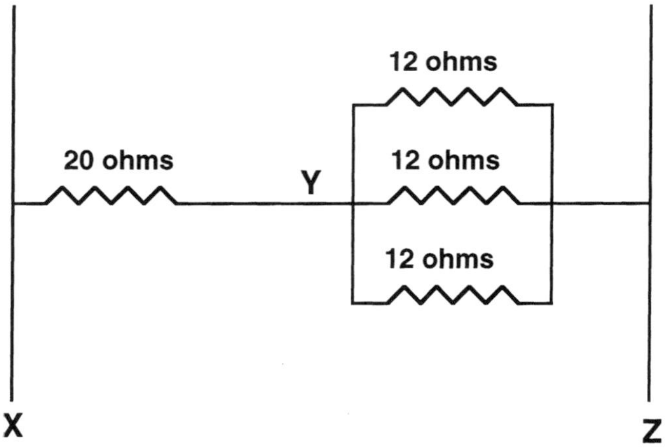

EACH OF THE 12 OHM LOADS IS 2 AMPERES

(a) 720 watts (b) 768 watts (c) 144 watts (d) 864 watts

18. The ungrounded conductor in a single-phase circuit can be determined by using ____.

(a) a manometer (b) a galvanometer (c) a neon tester (d) an ohmmeter

19. The decimal equivalent of 5/8" is _____.

(a) 0.652 (b) 0.500 (c) 0.875 (d) 0.625

20. Which of the following statements is correct?

(a) Four equipment grounding conductors in a box count as four conductors.
(b) The maximum overcurrent device for a #14 THW conductor is 20 amperes.
(c) Emergency lighting systems should be connected to two separate sources of power.
(d) A ground rod connection must be visible.

21. What Chapter in the Code addresses elevators?

(a) Chapter 5 (b) Chapter 6 (c) Chapter 7 (d) Chapter 8

22. The Code requires that a conduit with more than two conductors to be filled to a maximum of 40 percent. The reason for the limitation on the fill area is ____.

(a) to reduce the heat from induction
(b) to allow for future fill
(c) for better circulation of air
(d) to make the pulling of wires easier without abrasion

23. When applying rubber tape to a splice, the tape should be applied with ____.

(a) enough tension to pull the tape to about half its original width
(b) an adhesive first before applying the tape
(c) a heat gun
(d) only copper conductors

24. "Nichrome" wire is usually used for ____.

(a) heater coils (b) motor field windings (c) interpole windings (d) light bulb filaments

25. If a 240 volt electric heater is used on a 120 volt circuit the amount of heat produced will be ____.

(a) one half as much (b) twice as much (c) four times as much (d) one fourth as much

26. If a 10Ω, a 20Ω and a 30Ω resistor are connected in series across a 120 volt source, the voltage across the 20Ω resistor will be ____ volts.

(a) 20 (b) 40 (c) 60 (d) none of these

27. Which of the following is the symbol for wiring connected?

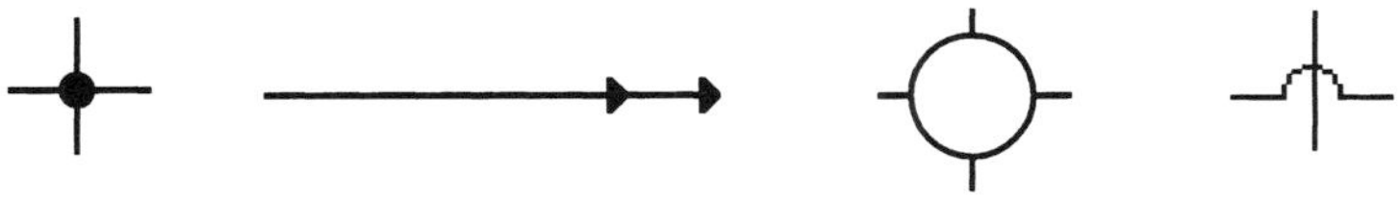

(a) (b) (c) (d)

28. A one-eighth bend in a conduit is equivalent to an angle of ____ degrees.

(a) 33 (b) 45 (c) 18 (d) 22

29. A 4Ω, a 6Ω, a 10Ω and a 15Ω are connected in parallel. Which resistor will consume the most power?

(a) 4Ω (b) 6Ω (c) 10Ω (d) 15Ω

30. The most common cause for a worker to lose balance and fall from a ladder is ____.

(a) exerting a heavy pull on an object which gives suddenly
(b) using a ladder that is too long
(c) to have too much weight in his tool belt
(d) slipping on the fiberglass rungs

31. Which one of the following is known as an actuating control?

(a) thermostat (b) relay (c) manometer (d) galvanometer

32. What is the total wattage of the four - 12Ω loads?

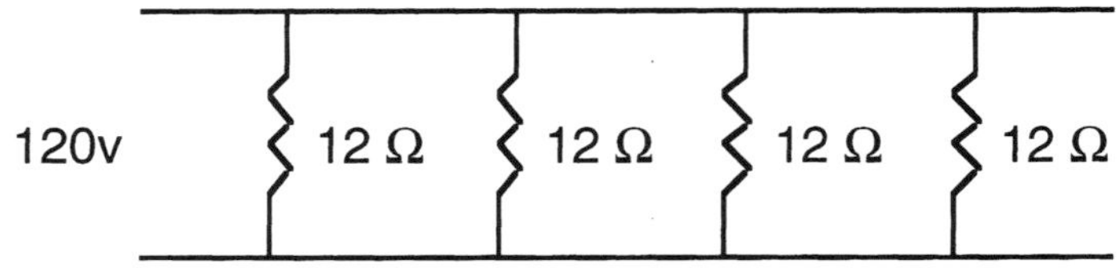

(a) 300 watts (b) 2400 watts (c) 4800 watts (d) none of these

33. A cycle counter would be used in testing ____.

(a) motors (b) transformers (c) ammeters (d) relays

34. With a given voltage, six light bulbs will consume the most wattage when connected _____.

(a) three in series and three in parallel (b) four in parallel and two in series
(c) all in parallel (d) all in series

35. Which of the following has the poorest conductance?

(a) aluminum (b) carbon (c) brass (d) copper

36. When determining the size of cubic inch for a device box, a duplex receptacle is counted as ____ wires.

(a) one (b) two (c) three (d) four

37. Which of the following plugs is a polarized plug?

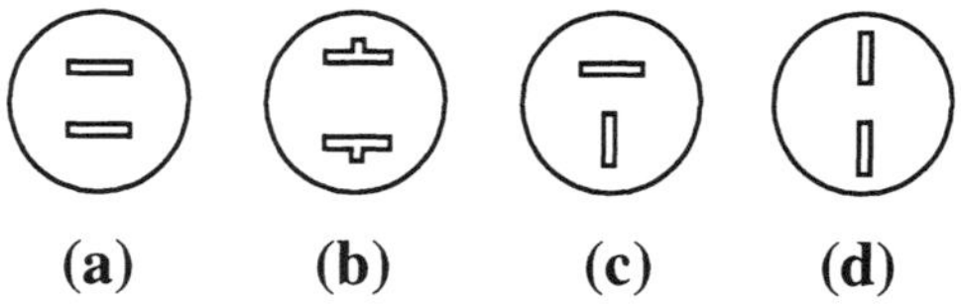

38. The electrolyte of a storage battery is formed by the dissolving of ____ in water.

(a) sulphuric acid (b) hydrochloric acid (c) lye (d) soda

39. The Code refers to low voltage as ____.

(a) 600 volts and under (b) 240 volts (c) 120 volts (d) 12 volts

40. Orangeburg pipe is ______.

(a) nonmetallic of fiber (b) cast iron (c) galvanized steel (d) lead coated

41. What is the power of the four - 12Ω loads?

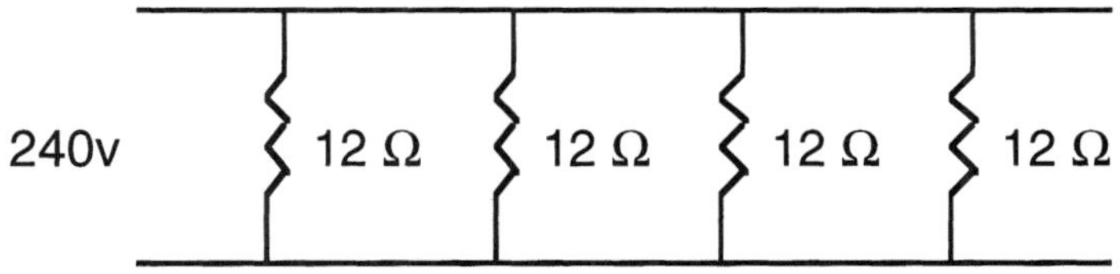

(a) 4800 watts (b) 9600 watts (c) 19,200 watts (d) none of these

42. Which of the following is the symbol for a temperature actuated switch?

(a) (b) (c) (d)

43. Which of the following will have the **least** effect on the voltage drop of a branch circuit?

(a) the size of the conductors
(b) the amount of the load
(c) whether the source is 50 hz or 60 hz
(d) the length of the conductors

44. A low value of reactive voltamperes in an AC circuit compared with the wattage would indicate _____.

(a) unity power factor
(b) high power factor
(c) maximum current for the load
(d) very low efficiency

45. It is generally recognized that _____ is the largest size EMT hand bender.

(a) 1" (b) 1 1/4" (c) 1 1/2" (d) 2"

46. Which of the following is a short circuit?

(a) a hot wire touching a grounded metal frame
(b) a short conductor with a high resistance
(c) arcing between a hot conductor and a metal box
(d) a very low resistance between the circuit conductors

47. When the current flowing in a circuit exceeds the allowable capacity of the conductor the part of the circuit that melts is called a _____.

(a) thermal overload (b) heater (c) breaker (d) fuse

48. If a light bulb rated 100 watts @ 120 volts is connected across a 240 volt source the wattage would be _____.

(a) 100 watts (b) 200 watts (c) 300 watts (d) 400 watts

49. The ratio of the power in kw to the total kva of an AC circuit is known as the _____.

(a) demand factor (b) diversity factor (c) power factor (d) correction factor

50. With only switch 4 closed and a line voltage of 225 volts, the drop across one of the 10 ohm resistors is _____ volts.

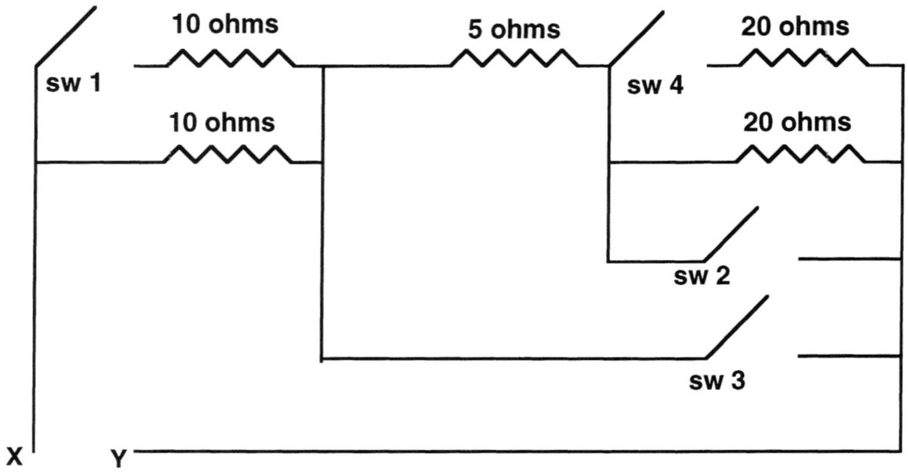

(a) 225 (b) 90 (c) 64.3 (d) 56.3

2005

CLOSED BOOK EXAM #14

50 QUESTIONS
TIME LIMIT - 1 HOURS

TIME SPENT ______ **MINUTES**

SCORE ______ **%**

TH

MASTER CLOSED BOOK EXAM #14 One Hour Time Limit

1. The National Electrical Code is sponsored by the ____.

(a) Underwriters Lab
(b) National Safety Council
(c) National Electrical Manufacturers Association
(d) National Fire Protection Association

2. A drawing showing the floor arrangement of a building is referred to as a(an) ____.

(a) perspective (b) isometric (c) surface G.B. (d) plan

3. Readily accessible is ____.

(a) requiring only a 6 foot ladder (b) requiring to move only small obstacles
(c) within 50 feet (d) capable of being reached quickly for operation

4. A fitting is ____.

(a) part of a wiring system that is intended primarily to perform an electrical function
(b) pulling cable into a confined area
(c) to be suitable or proper for
(d) part of a wiring system that is intended primarily to perform a mechanical function

5. The neutral conductor _____.

(a) is always the "white" grounded conductor (b) carries the balanced current
(c) Note 8 never applies (d) carries the unbalanced current

6. An oil switch is a switch having contacts which operate under ____.

I. askarel II. oil III. other suitable liquids

(a) I only (b) II only (c) I and II only (d) I, II or III

7. Which of the following is an ammeter?

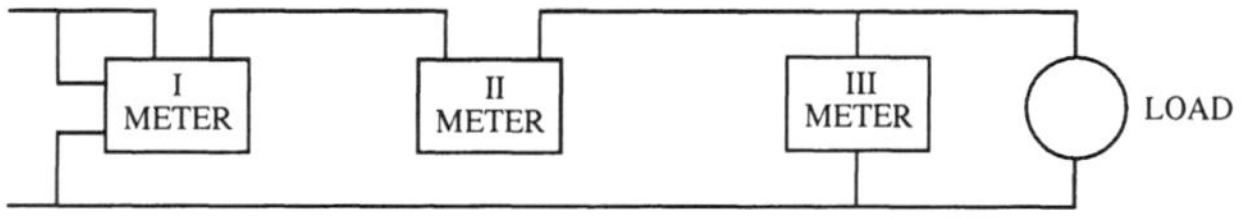

(a) I only (b) II only (c) III only (d) I, II or III

8. All wiring must be installed so that when completed ____.

(a) it meets the current-carrying requirements of the load
(b) it is free of shorts and unintentional grounds
(c) it is acceptable to Code compliance authorities
(d) it will withstand a hy-pot test

9. If you were installing a fluorescent fixture in a customer's home and the noise emitted by the ballast were an important consideration, which one of the following sound ratings would you select to give the quietest operation?

(a) Type A (b) Type D (c) Type F (d) Type AB

10. 60 cycle frequency travels 180 degrees in how many seconds?

(a) 1/60 (b) 1/120 (c) 1/180 (d) 1/30

11. A wattmeter indicates ____.

I. real power II. apparent power if PF is not in unity III. power factor

(a) I only (b) II only (c) III only (d) I, II and III

12. Of the following, improving the commutation of a DC generator is most often done by means of ____.

(a) a compensator **(b) interpoles**
(c) an equalizer **(d) a series rheostat in series with the equalizer**

13. Code requirements may be waived ____.

(a) by the Fire department **(b) by the utility company**
(c) by the authority having jurisdiction **(d) under no circumstances**

14. Ampacity is ____.

(a) the current-carrying capacity of conductors expressed in volt-amps.
(b) the current-carrying capacity expressed in amperes.
(c) the current-carrying capacity of conductors expressed in amperes.
(d) the current in amperes a conductor can carry continuously under the conditions of use without exceeding its temperature rating.

15. An electric circuit that controls another circuit through a relay is a ____ circuit.

(a) remote-control (b) pilot (c) low-energy power (d) transfer

16. Concrete, brick or tile walls are considered as being ____.

(a) isolated (b) insulators (c) grounded (d) dry locations

17. Which of the following systems will furnish single-phase, three-phase, and two voltage levels?

(a) three-phase, 3-wire **(b) three-phase, 4-wire**
(c) two-phase, 3-wire **(d) single-phase, 4-wire**

18. All edges that are invisible should be represented in a drawing by lines that are ____.

(a) dotted (b) curved (c) solid (d) broken

19. The definition of kcmil is ____.

(a) correction factor (b) one million (c) one-thousand circular mils (d) one-hundred mils

20. An appliance that is not easily moved from one place to another in normal use is a ____ appliance.

(a) fastened in place (b) dwelling-unit (c) fixed (d) stationary

21. A point on the wiring system at which current is taken to supply utilization equipment is called ____.

(a) a receptacle (b) a junction (c) an outlet (d) a tap

22. ____ is a protective device for assembly as an integral part of a motor or motor-compressor and which, when properly applied, protects the motor against dangerous overheating due to overload and failure to start.

(a) Thermal protector **(b) Thermal cutout**
(c) Thermal overcurrent **(d) Thermal heat anticipator**

23. Listed equipment must be installed or used in accordance with ____.

(a) any instructions included in listing or labeling **(b) the blue print**
(c) instructions provided by the manufacturer of the equipment **(d) the electrician**

24. Compliance with the provisions of the Code will result in ____.

(a) freedom from hazard **(b) an efficient system**
(c) good electrical service **(d) convenient operation**

25. The horsepower rating of a motor ____.

(a) is a measure of motor efficiency **(b) is the input to the motor**
(c) cannot be changed to watts **(d) is the output of the motor**

26. What is the function of a neon glow tester?

I. determines if circuit is alive
II. determines polarity of DC circuits
III. determines if circuit is AC or DC current

(a) I only (b) II only (c) III only (d) I, II and III

27. Power in a three-phase system may be measured with a minimum of ____.

(a) one wattmeter (b) two voltmeters (c) two ammeters (d) none of these

28. What are Optical Fiber Cables used for?

I. transmit light for control II. signaling III. communications

(a) I only (b) II only (c) III only (d) I, II and III

29. What Chapter in the Code is power limited circuits referred to?

(a) Chapter 5 (b) Chapter 6 (c) Chapter 7 (d) Chapter 8

30. Something that would effect the ampacity of a conductor would be ____.

I. voltage II. amperage III. length IV. temperature

(a) I only (b) II only (c) III only (d) IV only

31. An electric bell outfit would be used to check for ____.

(a) voltage (b) ampacity (c) continuity (d) current

32. An ammeter is connected ____.

(a) in series with the load
(b) in parallel with the load
(c) both in series and in parallel with the load
(d) neither in series nor in parallel with the load

33. The Code requires all circuits to have at least ____ and adequate grounding.

(a) an ungrounded wire and a grounding wire **(b) two hot wires and a grounding wire**
(c) at least two wires **(d) a hot wire and a grounding wire**

34. Which of the following is **not** permitted for connecting conductors #6 and larger to terminal parts?

(a) solder lugs (b) pressure connectors (c) splices to flexible leads (d) wire binding screws

35. ____ are used to provide explanatory material, references to other sections, precautions to be observed, and further mandatory requirements.

(a) Mandatory rules (b) Exceptions (c) Fine print notes (d) Listings

36. Accessible (as applied to wiring methods) is ____.

I. not permanently closed in by the structure or finish of the building.
II. capable of being removed or exposed without damaging the building structure or finish.

(a) I only (b) II only (c) both I and II (d) neither I nor II

37. A transformer would most likely have a ____ efficiency.

(a) 60% (b) 70% (c) 80% (d) 90%

38. Continuous duty is ____.

(a) a load where the maximum current is expected to continue for three hours or more.
(b) a load where the maximum current is expected to continue for one hour or more.
(c) intermittent operation in which the load conditions are regularly recurrent.
(d) operation at a substantially constant load for an indefinitely long time.

39. The definition of dustproof is ____.

(a) an electrical enclosure with sealing gaskets to prevent the equipment from overheating from
an accumulation of dust.
(b) constructed so that dust will not enter the enclosing case under specified test conditions.
(c) protected so that dust will not interfere with its successful operation.
(d) a housing that seals dust from contacting energized parts.

40. It is generally not good practice to supply lights and motors from the same circuit because ___.

(a) lamps for satisfactory service must operate within closer voltage limits than motors.
(b) overloads and short circuits are more common on motor circuits.
(c) when motors are started, the large starting current causes a voltage drop on the circuit and the lights will blink or burn dim
(d) all of these

41. Covered, shielded, fenced or enclosed by means of suitable covers, casings, barriers, rails, screens, mats, or platforms is the definition of ____.

(a) guarded (b) protected (c) isolated (d) enclosed

42. The overhead service conductors from the last pole or other aerial support to and including the splices, if any, connecting to the service entrance conductors at the building or other structure is the ____.

(a) service load (b) service lateral (c) service drop (d) service supply

43. Without live parts exposed to a person on the operating side of the equipment is called ____.

(a) dead front (b) isolated (c) externally operable (d) interrupted

44. Frequency is **not** a problem in most electrical work because it ____.

(a) never affects circuit values (b) varies directly with voltage
(c) is a constant value (d) only concerns DC

45. If the voltage is doubled, the ampacity of a conductor ____.

(a) increases (b) decreases (c) doubles (d) remains the same

46. To calculate the power used by an appliance (not motor driven) one needs to know ____.

I. voltage and current II. current and resistance III. voltage and resistance

(a) I only (b) II only (c) III only (d) I, II or III

47. Fixture wire shall not be smaller than # ____.

(a) 16 (b) 18 (c) 20 (d) 22

48. A toaster will produce less heat on low voltage because ____.

(a) its total watt output decreases
(b) the current increases
(c) the resistance has changed
(d) its total watt output increases

49. Electrical equipment shall be installed ____.

(a) better than the minimum Code allows
(b) according to the local Code when more stringent than the N.E.C.
(c) according to the N.E.C. regardless of the local Code
(d) according to the local Code when less stringent than the N.E.C.

50. What is the applied voltage of this circuit?

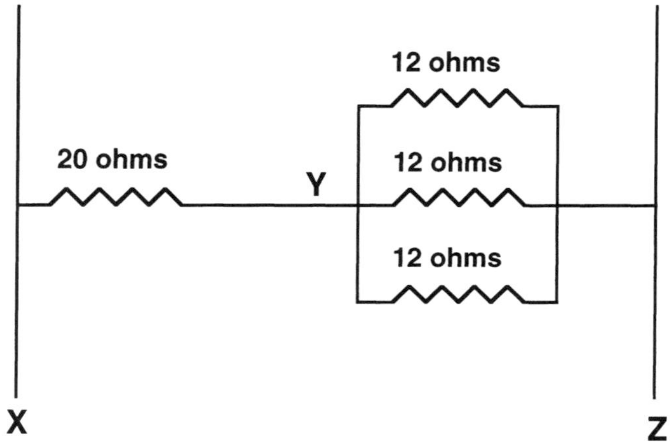

EACH OF THE 12 OHM LOADS IS 2 AMPERES

(a) 144 volts (b) 120 volts (c) 784 volts (d) 336 volts

2005

OPEN BOOK EXAM #1

70 QUESTIONS
TIME LIMIT - 2 HOURS

TIME SPENT ______ **MINUTES**

SCORE ______ **%**

MASTER OPEN BOOK EXAM #1 Two Hour Time Limit

1. What **MINIMUM** width clearance is required in front of enclosed 240 volt equipment which encorporates 24" hinged access panels?

(a) 16" (b) 24" (c) 30" (d) 36"

2. The grounding electrode system of each building or structure served shall be comprised of all of the following **EXCEPT** ____.

(a) a concrete encased electrode (b) an 8' long aluminum pipe electrode
(c) a metal frame of the building (d) a metal underground water pipe

3. What is the **MINIMUM** size copper grounding conductor required to serve a multisection motor control center equipped with a 300 amp overcurrent device?

(a) #6 (b) #4 (c) #3 (d) #2

4. What is the **MAXIMUM** permitted load rating on branch circuits supplying signs or outline lighting systems using neon tubing?

(a) 15 amps (b) 20 amps (c) 25 amps (d) 30 amps

5. Which of these motors can be protected against overload by a single overload unit?

(a) Three phase AC
(b) Single phase AC with three phase supply
(c) Two phase AC with 5 wire, two phase AC and grounded neutral
(d) Two phase AC with 3 wire, two phase AC supply and one grounded conductor

6. What is an approved location for the installation of low voltage lighting track operating at 24 volts?

(a) A car wash final rinse spray area
(b) In a protected display area less than 5' above the floor
(c) In a storage battery room above the batteries
(d) A pool supply liquid chlorine re-filling station

7. What is the **MINIMUM** size underground service lateral for a copper-clad aluminum single-branch circuit serving a controlled water heater?

(a) #14 (b) #12 (c) #10 (d) #8

8. What is the **MINIMUM** size copper equipment grounding conductor which may be used for a circuit wired with size #2/0 copper THW and protected by 125 amp circuit breaker?

(a) #6 (b) #4 (c) #2 (d) #2/0

9. Where are conduit seals **NOT** required in a Class I Division 1 installation?

(a) Where the conduit exits the Class I Division 1 area
(b) Where the conduit enters an explosion-proof motor
(c) Where a conduit less than 36" in length connects two enclosures
(d) Where metal conduit passes completely through the Class I Division 1 area with no fittings less than 12" outside any classified area

10. Disregarding exceptions, if the copper, ungrounded conductors of a 120/240 volt, single-phase dwelling service are #3/0, what is the **MINIMUM** allowable size for the copper grounding electrode conductor?

(a) #4 (b) #2 (c) #1 (d) #1/0

11. What is the **MAXIMUM** allowable voltage that conductors within electrical nonmetallic tubing may carry?

(a) 150v (b) 300v (c) 600v (d) 1000v

12. Where shall mobile home service equipment be located?

(a) Adjacent to the mobile home being served
(b) Within sight and not more than 50' from the mobile home served
(c) In a mobile home without permission of the authority having jurisdiction
(d) On the mobile home without permission of the authority having jurisdiction

13. Aluminum service entrance phase conductors are larger than #1750 kcmil. The bonding jumper shall have an area of not less than what percentage of the area of the largest phase conductor?

(a) 12.5% (b) 70% (c) 80% (d) 125%

14. How much space is occupied by a #350 kcmil RHW grounding conductor without an outer covering?

(a) 0.5958 sq.in. (b) 0.7870 sq.in. (c) 0.8710 sq.in. (d) 1.0010 sq.in.

15. What is the **MAXIMUM** support spacing for #4/0 copper conductors in vertical raceways?

(a) 60' (b) 80' (c) 100' (d) 200'

16. When an aluminum conductor is used outside for the grounding electrode conductor, what is the **MINIMUM** number of inches that it must be from earth?

(a) 6" (b) 12" (c) 18" (d) 24"

17. What is the classification of a location where ignitable concentrations of flammable gas are likely to be present during normal operations?

(a) Class I, Division 1 (b) Class I, Division 2 (c) Class II, Division 1 (d) Class III, Division 1

18. What is the **MINIMUM** number of feet that open conductors rated at 480 volts shall be above a commercial driveway which is subject to truck traffic?

(a) 8' (b) 16' (c) 18' (d) 22'

19. A chemical blending operation using acetone and alcohol falls into which hazardous location classification?

(a) Class I (b) Class II (c) Class III (d) Class IV

20. How many light fixtures may be permanently attached to a residential small appliance branch circuit?

(a) One (b) Two (c) Three (d) None

21. What is the **MINIMUM** depth for direct burial of 24 volt Type UF landscape lighting cable?

(a) 6" (b) 12" (c) 18" (d) 24"

22. Each resistance welder shall have overcurrent primary protection set at not more than ____ percent.

(a) 200 (b) 300 (c) 250 (d) 125

23. Type USE-2 is rated at ____.

(a) 60°C (b) 75°C (c) 90°C (d) 86°F

24. When the letters THHN are stamped on the insulation of a conductor, the HH indicates that the conductor insulation has a temperature rating of ___ degrees C.

(a) 60 (b) 75 (c) 90 (d) 110

25. If a nipple 18" long contains 24 conductors, the ampacity for each conductor must be reduced to ____ of Table 310.16 and Table 310.18.

(a) 80% (b) 70% (c) 60% (d) 0%

26. The rated primary current of an electric resistance welder is ____.

I. 2.8 times the primary current divided by the duty cycle
II. determined by the percentage of time during which the welder is loaded
III. the rated kva multiplied by 1000 and divided by the rated primary voltage, using values given on the nameplate
IV. the current drawn from the supply circuit during each welder operation at the particular heat tap and control setting used

(a) I and III only (b) II and IV only (c) III only (d) I, II, III and IV

27. Recreational vehicles and parks grounding terminals shall be of the solderless type and listed as pressure terminal connectors recognized for the wire size used. The bonding conductor shall be solid or stranded, insulated or bare, and shall be #____ AWG copper minimum, or equal.

(a) 10 (b) 8 (c) 6 (d) 4

28. External surfaces of heating equipment that exceed _____ shall be guarded, isolated, or thermally insulated to protect against contact by personnel in the area.

(a) 120 degrees C (b) 140 degrees C (c) 140 degrees F (d) 86 degrees F

29. Where lamps are of a size or type that may, under normal operating condition, reach surface temperatures exceeding ___ of the ignition temperature in degrees Celsius of the gas or vapor involved, shall comply with the requirements of the N. E. C. for Class I, Division I or shall be of a type that has been tested in order to determine the marked operating temperature or temperature class (T Code).

(a) 50% (b) 75% (c) 80% (d) 100%

30. Signs and outline lighting equipment used in wet locations shall comply with which of the following conditions ___ before the drain holes are suitable for wet conditions.

(a) Drain holes shall not be larger than 26 mm (1 in.) or smaller than 12 mm (1 /2 in.).
(b) Every low point or isolated section of the equipment shall have at least two drain holes.
(c) Drain holes shall be positioned such that there will be no external obstructions.
(d) All of the above

31. Over 600v, where fuses are subject to being energized by a backfeed, a sign shall be placed on the enclosure ___ identifying this hazard.

(a) back panel (b) fuse (c) door (d) wiring

32. The pipeline, vessel, or both, being heated that is operating at a voltage greater than 30 but not more than 80 shall be ______ at designated points.

(a) grounded (b) de-energized (c) ungrounded (d) connected

33. Fire Alarm circuits are classified as _____.

(a) restricted (b) power limited (c) non power limited (d) both (b) and (c)

34. The process of supplying an enclosure with a protective gas with or without continuous flow at ____ pressure to prevent the entrance of combustible dust, or an ignitible fiber or flying.

(a) 50 pounds (b) 35 pounds (c) sufficient (d) low

35. A lighting system limited to 30 volts or less is operate at no more than ___ amperes maximum supplying lights.

(a) 10 (b) 15 (c) 20 (d) 25

36. The Code states there shall be a circuit directory located at the ______ .

(a) back of the panel
(b) inside or outside the panel door
(c) wall adjacent to the panel
(d) only required on the plans

37. The largest rigid non-metallic conduit permitted to be used is _____ inches.

(a) 3 (b) 4 (c) 5 (d) 6

38. When installing a fire alarm system __ cable is to be used to ensure continued operation of the system for a specific time under fire conditions.

(a) FA (b) FM (c) NM (d) CI

39. The floors of vaults in contact with the earth shall be of concrete that is not less than __ inches thick.

(a) 3.5 (b) 4 (c) 5 (d) 6

40. When feeders supplying transformers the ampacity of feeder conductors shall not be _____ than the sum of the nameplate ratings of the transformers supplied when only transformers are supplied.

(a) less (b) more (c) equal (d) greater

41. Materials used for floors heated in excess of 30°C or ____ shall be identified for use at these temperatures.

(a) 76°F (b) 80°F (c) 86°F (d) 90°F

42. The smallest rigid metallic conduit permitted to be used is _____ inch.

(a) 1/4 (b) 1/2 (c) 3/4 (d) 5/8

43. In a single family dwelling where a CATV cable is stapled alongside floor joists, it is allowable to run the cable through an air plenum?

(a) There are no restrictions on low voltage wiring.
(b) Yes, but only when the plenum temperature does not exceed 36°C.
(c) The cable must be stapled 18" away from the plenum.
(d) Yes, if it is type CATVP cable.

44. In a swimming pool, underwater lighting fixtures supplied by a 120 volt branch circuit shall be protected by a ground-fault circuit-interrupter if the fixtures operate at more than ____ volts.

(a) 10 (b) 12 (c) 15 (d) 24

45. A point located 24" above grade level and 20 feet from the edge of an indoor remote gas pump is considered ____.

(a) Class I, Group D, Division 2 (b) Class I, Group D, Division 1
(c) Class I, Group C, Division 2 (d) Class I, Group C, Division 1

46. In commercial garages using electrical hand tools, portable lights, etc., ground fault protection shall be provided for _____.

(a) receptacles in pits below floor level only
(b) receptacles located in adjacent bathrooms only
(c) receptacles within 18" above the floor only
(d) all receptacles in the service area

47. In an open paint spraying area, the Division 2 area extends ____ horizontally outside of the Division 1 location.

(a) 8' (b) 15' (c) 20' (d) unlimited

48. A propane-dispensing unit is located outdoors, 50 feet from an office in which the branch circuit supplying the unit originates. Conduit seals shall be required ____.

(a) where the conduit emerges from the ground at the office only
(b) where the conduit emerges from the ground at the dispensing unit only
(c) where the conduit emerges from the ground at both the dispensing unit and at the office
(d) No seals are required for propane gas

49. In a hospital electrical system, the Critical Branch of the Emergency System shall supply power for ____.

(a) coronary care units (b) human physiology labs
(c) anesthetizing locations (d) all of these

50. On a 15 or 20 ampere branch circuit, the total rating of appliances shall not exceed ____.

(a) 50% of the branch circuit rating
(b) 80% of the branch circuit rating
(c) 50% of the branch circuit rating when lighting units, cord-and-plug connected appliances not fastened in place, or both, are also supplied
(d) 80% of the branch circuit rating when lighting units or portable or stationary appliances are also supplied

51. The grounded conductor at a service entrance shall not be smaller than the required size of the ____ conductor.

(a) phase (b) service (c) equipment (d) grounding electrode

52. If one ground rod is installed and the resistance to ground is measured to be 50 ohms, the National Electrical Code requires ____.

(a) installation of one additional ground rod
(b) relocation of the ground rod to meet the 25 ohm maximum requirement
(c) increasing the diameter and/or length of the rod to meet the 25 ohm maximum requirement
(d) installation of one or more additional ground rods until the 25 ohm maximum requirement is met

53. Type JM or SM cable may be run through an environmental return air duct when enclosed in ____.

(a) metal raceways (b) PVC (c) ENT (d) flexible nonmetallic tubing

54. In a fountain, flexible cord immersed in or exposed to water shall be of the ____ type , as designated in Table 400.4 and shall be listed with a "W" suffix.

(a) Thermoplastic elastomer (b) Thermoset (c) Submersible (d) Extra hard usage

55. The purpose of transfer equipment as used in emergency systems is ____.

(a) a means of opening and closing the ungrounded conductors
(b) to connect emergency power even though normal power is still available
(c) a means of opening and closing both the grounded and ungrounded conductors
(d) to prevent inadvertent parallel interconnection of normal and emergency supply sources

56. Disregarding exceptions, the minimum cover requirement for rigid nonmetallic conduit approved for direct burial with a circuit voltage of 25,000 volts is ____.

(a) 18" (b) 30" (c) 2' (d) 4'

57. Fire alarm signal cables, when run exposed on walls or ceilings, shall be securely fastened at intervals of not more than 18" where located within ____ of the floor.

(a) 30" (b) 4' (c) 7' (d) 7 1/2'

58. The requirements of the National Electrical Code may be waived by the authority having jurisdiction ____.

(a) where it is assured that equivalent objectives can be achieved
(b) only with approval of the State Licensing Board
(c) only if replaced with standards which are more stringent
(d) only if approved by the NFPA

59. The headroom required for a motor control center rated 240 volts is ____.

(a) 6' (b) 6' 6" (c) 7' (d) 7' 6"

60. No grounded conductor shall be attached to ____.

(a) any terminal or lead so as to reverse designated polarity
(b) the screw shell of a lampholder or other device
(c) a terminal unless identified by a green color
(d) none of the above

61. Of the following, the only additional load permitted on a residential gas-fired furnace branch circuit is a _____.

(a) humidifier
(b) service receptacle
(c) sump pump within the same room
(d) lighting outlet in furnace room

62. If a subpanel is added to an existing system with a main panel, the subpanel shall be grounded by _____.

(a) connecting the main and subpanel with a bonding jumper
(b) installing a separate grounding electrode and grounding electrode conductor
(c) installing a separate grounding electrode conductor to the existing building grounding electrode
(d) running a separate equipment grounding conductor in the same raceway with the circuit (feeder) conductors back to the main panel

63. Flexible cords may be used in all of the following locations except _____.

(a) on hoists
(b) on cranes
(c) as pendants
(d) to substitute for fixed wiring

64. An outlet box for the tap conductor of a recess fixture terminal connection shall be placed at a minimum distance of _____ from the fixture.

(a) 6"   (b) 12"   (c) 18"   (d) no specific requirement by the Code

65. A motor controller enclosure intended for general purpose indoor use wherever atmospheric conditions are normal is _____.

(a) NEMA type 1   (b) NEMA type 3R   (c) NEMA type 6   (d) NEMA type 13

66. Several motors, each not exceeding one horsepower in rating, shall be permitted on a nominal 120 volt branch circuit protected at not over 20 amperes or a branch circuit of 600 volts, nominal, or less, protected at not over 15 amperes, if _____.

(a) individual overload protection confirms to 430.32
(b) the full-load rating of each motor does not exceed 6 amperes
(c) the rating of the branch-circuit short-circuit and ground fault protective device marked on any of the controllers is not exceeded
(d) All of the above

67. Where a surface mounted light fixture is installed on the ceiling of a clothes closet, which of the following is not permitted by the Code?

(a) Recessed fluorescent fixture.
(b) Totally enclosed incandescent lamp.
(c) Incandescent lamp supported by a cord.
(d) Surface mounted fluorescent fixture.

68. The ampacity of phase conductors from generator terminals to the first overcurrent device shall be not less than ____ of the nameplate current rating of the generator.

(a) 75% (b) 80% (c) 115% (d) 125%

69. A room air conditioner shall be considered as a single motor unit in determining its branch circuit requirements when which of the following conditions is/are met?

(a) The total rated-load current is shown on the room air conditioner nameplate.
(b) It is rated not more than 40 amps and 250 volts, single phase.
(c) It is cord-and-attachment plug-connected.
(d) All of the above.

70. Fixed wiring, which is to provide external power to aircraft hangers, shall be installed at least ____ above floor level.

(a) 12" (b) 18" (c) 24" (d) 30"

2005

OPEN BOOK EXAM #2

70 QUESTIONS
TIME LIMIT - 2 HOURS

TIME SPENT [ ] **MINUTES**

SCORE [ ] **%**

MASTER OPEN BOOK EXAM #2 **Two Hour Time Limit**

1. What is the **MINIMUM** size wire permitted as a bonding jumper between the communications grounding electrodes and power grounding electrodes in a communications circuit?

(a) #14 (b) #12 (c) #6 (d) This connection is not permitted.

2. What is the **MAXIMUM** percentage of rated primary current that can be used to protect an arc welder from overcurrent?

(a) 125% (b) 150% (c) 175% (d) 200%

3. The calculated value for a motor branch-circuit short-circuit protective device does not correspond to a standard size. The rating of the device shall be ____.

(a) increased by installing larger overloads
(b) custom made to meet the rating
(c) increased to the next higher standard size
(d) decreased to the next lower standard size

4. Chain supported light fixtures used in a show window shall be permitted to be ____.

(a) hung with 50# chain
(b) less than 6' above the floor line
(c) externally wired
(d) none of these

5. What is the **MAXIMUM** permitted rating on branch circuits supplying signs or outline lighting systems?

(a) 15 amps (b) 20 amps (c) 25 amps (d) 30 amps

6. It shall be permissible to extend ____ of surface metal raceways through dry walls, dry partitions, and dry floors.

(a) 2 feet (b) 3 feet (c) 6 feet (d) unbroken lengths

7. Any room or location in which flammable anesthetics are stored shall be considered to be a ____ location from floor to ceiling.

(a) Class I, Division 1 (b) Class I, Division 2 (c) Class II, Division I2 (d) nonhazardous

8. Enclosures for isolating switches in Class I, Division II locations ____.

(a) may be of general-purpose type
(b) must be explosion-proof type
(c) must have interlocking devices
(d) may not have doors in hazardous area

9. Electric-discharge tubing operating at more than 7500 volts shall clear the adjacent surface by not less than ____.

(a) 1/4"   (b) 1/2"   (c) 1"   (d) 3"

10. Conductors to the hoistway door interlocks from the hoistway riser shall be ____.

(a) flame retardant   (b) type SF or equivalent
(c) rated 200 degrees C   (d) all of these

11. The Code provides that in Class I, Division 1 locations conduit seals shall be placed not farther from spark-producing devices than ____ inches.

(a) 18   (b) 24   (c) 12   (d) 30

12. Steel cable trays shall not be used as equipment grounding conductors for circuits protected above ____ amperes.

(a) 200   (b) 60   (c) 600   (d) 1200

13. Which of the following are **not** classified patient care areas?

I. day rooms   II. lounges   III. business offices

(a) II only   (b) II and III only   (c) III only   (d) I, II and III

14. Which of the following applies to Class I Division I locations?

(a) ignitible flammable gases and vapors   (b) grain silos
(c) ignitible fibers or flyings   (d) combustible dust

15. For straight pulls, the length of the box shall be not less than ____ the outside diameter, over sheath, of the largest conductor or cable entering the box on systems over 600 volts.

(a) 8 times   (b) 6 times   (c) 36 times   (d) 48 times

16. Each circuit leading to or through a dispensing pump shall be provided with a switch or other acceptable means to disconnect____from the source of supply all conductors of the circuit, including the grounded conductor, if any.

(a) automatically   (b) simultaneously   (c) manually   (d) individually

17. Where storage batteries are used for emergency systems they shall be ____.

(a) provided with automatic battery charging means **(b) alkali or acid type**
(c) neither (a) nor (b) **(d) both (a) and (b)**

18. In a commercial garage the pit shall be classified ____ unless provisions are made for six air changes per hour.

(a) Class I, Division 2 (b) Class II, Division 2 (c) Class II, Division 1 (d) Class I, Division 1

19. No grounded conductor shall be attached to any terminal or lead so as to reverse designated ____.

(a) phase (b) angle (c) polarity (d) line

20. For general wiring in Class I, Division 1 locations it is permissible to use ____.

(a) rigid metal conduit (b) EMT (c) flexible metal conduit (d) all of these

21. Sealing compound is employed with mineral-insulated cable in a Class I location for the purpose of ____.

(a) preventing passage of gas or vapor **(b) excluding moisture**
(c) limiting a possible explosion **(d) preventing escape of powder**

22. All buildings or portions of buildings or structures designed or intended as a place of assembly shall have ____ or more persons.

(a) 50 (b) 100 (c) 250 (d) 500

23. A stage switchboard that is not completely enclosed dead-front and dead-rear or recessed into a wall shall be provided with a ____ extending the full length of the board to protect all equipment on the board from falling objects.

(a) cover (b) guard (c) mesh net (d) metal hood

24. Signs in wet locations shall be weatherproof and ____.

I. have drain holes positioned so there is no external obstructions
II. have at least one drain hole in every low point
III. drain holes shall not be smaller than 1/4"

(a) I and II only (b) II and III only (c) I only (d) I, II and III

25. Where magnesium, aluminum or aluminum-bronze powders may be present, transformers ____.

(a) must be dust-tight **(b) may be pipe-ventilated**
(c) must be approved for Class II, Division 1 **(d) are not allowed**

26. The conductor between a lightning arrester and the line for installations operating at 1000 volts or more must be at least ____ copper.

(a) #14 (b) #6 (c) #8 (d) #12

27. A park trailer is one built on a single chassis mounted on wheels and having a gross trailer area not exceeding ____ square feet in the set-up mode.

(a) 400 (b) 450 (c) 500 (d) 600

28. Health care low voltage equipment frequently in contact with bodies of persons shall not exceed ____ volts.

(a) 50 (b) 115 (c) 10 (d) 8

29. Color coding shall be permitted to identify ____ conductors where they are colored light blue and where no other conductors colored light blue are used.

(a) fire alarm (b) elevator (c) intrinsically safe (d) electrolytic cell

30. Where extensive metal in or on buildings may become energized and is subject to personal contact ____ will provide additional safety.

(a) adequate bonding and grounding **(b) bonding**
(c) suitable ground detectors **(d) none of these**

31. In general, Class 2 control circuits and power circuits ____.

I. may occupy the same raceway II. shall be installed in different raceways

(a) I only (b) II only (c) I and II (d) none of these

32. DC conductors used for electroplating shall be protected from overcurrent by ____.

I. a current sensing device which operates a disconnecting means
II. fuses or circuit breakers
III. other approved means

(a) II only (b) I and II only (c) II and III only (d) I, II and III

33. Where rigid PVC conduit is used as a raceway system in bulk storage plant wiring, the raceway shall include ____.

(a) sunlight resistant listing **(b) an equipment grounding conductor**
(c) a bushing with double locknuts **(d) PVC raceway is not permitted**

34. A ____ is a tank or vat in which electrochemical reactions are caused by applying electrical energy for the purpose of refining or producing usable materials.

(a) electrolytic cell (b) heating boiler (c) radiant heat (d) duct heater

35. Class 1 circuits and power supply circuits shall be permitted to occupy the same raceway only where the equipment powered is ____.

(a) low voltage (b) a fire alarm system (c) functionally associated (d) AC/DC

36. When a diesel engine is used as the prime mover of a generator to supply emergency power, how much of site fuel is required?

(a) one-half hour of fuel supply **(b) one hour of fuel supply**
(c) two hours of fuel supply **(d) three hours of fuel supply**

37. Non-shielded high-voltage cables shall be installed in ____ conduit encased in not less than 3" of concrete.

I. rigid PVC II. IMC III. rigid metal

(a) I only (b) II only (c) III only (d) I, II or III

38. The branch circuit overcurrent devices in emergency circuits shall be ____.

(a) of the reset type only **(b) a slow-blow type**
(c) accessible to only authorized personnel **(d) painted yellow**

39. _____ is the wall and arch that separates the stage from the auditorium.

(a) Stage drop (b) Proscenium (c) Acrolein (d) Butadiene

40. A manufactured wiring system shall have receptacles that are ____.

I. uniquely polarized II. of the locking type III. GFCI

(a) I only (b) II only (c) I and II only (d) I, II and III

41. Feeders to floating dwellings must be enclosed within ____ conduit in order to withstand the forces exerted by waves and tides.

(a) rigid metal (b) rigid PVC (c) liquidtight flexible (d) EMT

42. The heating of conductors depends on ____ values which, with generator field control, are reflected by the nameplate current rating of the motor-generator set driving motor rather than by the rating of the elevator motor, which represents actual but short-time and intermittent F.L.C. values.

(a) root-mean-square current **(b) voltage**
(c) ambient temperature **(d) ozone**

43. When installing office furnishings, receptacle outlets, ____ be located in lighting accessories.

(a) single-type only can **(b) duplex-type only can**
(c) shall **(d) shall not**

44. All of the following motors are permitted in a Class III, Division 1 area except ____.

(a) totally enclosed pipe ventilated **(b) non-ventilated**
(c) totally enclosed fan cooled **(d) water cooled**

45. In a Class 1 circuit, overcurrent protection for #16 conductors shall not exceed ___.

(a) 7 amperes (b) 10 amperes (c) 15 amperes (d) 20 amperes

46. Where low voltage fire alarm system cables penetrate a fire resistance rated wall, ___.

(a) no special considerations are required.
(b) junction box on each side of wall shall be metal conduit, nipple etc.
(c) openings shall be firestopped.
(d) cables cannot penetrate through a firewall.

47. Optical fiber cables are grouped into three types. Which of the following is not a type of optical cable?

(a) conductive (b) non-conductive (c) carrier (d) composite

48. A communication circuit run underground in a raceway, handhole, or manhole containing electric light or power conductors shall be in a section separated from such conductors by means of ___.

(a) metallic partitions **(b) non metallic conduit or guttering**
(c) not less than 12 inches of clear air space **(d) concrete**

49. A grounding conductor for a communications system shall not be smaller than ___.

(a) #8 cu (b) #10 cu (c) #12 cu (d) #14 cu

50. Sealed cans of gasoline may be stored in what type of area?

(a) Class I, Division 1 (b) Class I, Division 2 (c) Class II, Division 1 (d) Class II Division 2

51. If a remote control circuit for safety control equipment would introduce a direct fire or life hazard if their equipment should fail, this is considered a Class ___ circuit by the NEC.

(a) 1 (b) 2 (c) 3 (d) 4

52. In ducts or plenums used for environmental air, ___ may be used.

(a) type NM cable (b) insulated cable (c) type MI cable (d) type AC cable

53. Where 14 conductors are run in an 80 foot length of conduit, the allowable ampacity shall be reduced by a factor of ___ percent.

(a) 20 (b) 30 (c) 40 (d) 50

54. The maximum allowable voltage permitted for Class 1 remote control and signaling circuits is ___ volts.

(a) 30 (b) 150 (c) 300 (d) 600

55. An equipment grounding conductor run with circuit conductors in a raceway or cable ____.

(a) must be stranded wire
(b) may be bare or insulated
(c) must be insulated and have a continuous outer finish that is green
(d) may be black with green tape if smaller than a #6 conductor

56. The cross-sectional area of the conductors permitted in a seal shall not exceed ____ percent of the cross-sectional area of a rigid metal conduit of the same trade size unless it is specifically identified for a higher percentage of fill.

(a) 25% (b) 30% (c) 45% (d) 50%

57. A Class 1 power limited circuit shall be supplied by a source having a rated output not more than ___ volts.

(a) 12 (b) 15 (c) 24 (d) 30

58. Aircraft hangar attachment plugs and receptacles in Class I locations shall be designed so they cannot be energized while ____.

(a) the aircraft is within 20' **(b) the connections are being made or broken**
(c) the voltage is above 30 **(c) all of these**

59. Audio systems supplied by branch circuit power shall be protected by a ____ when used within 1.5 m of a pool, spa, hot tub, or fountain, or within 1.5 m of the prevailing or tidal high water mark.

(a) breaker (b) GFI (c) FRN fuse (d) REN fuse

60. Fire Alarm circuits shall ____.

(a) beinstalled to conform with 300.4(D **(b) have cables attached by staples or similar**
(c)be installed in a neat manner **(d) all of these**

61. Where flexible metal conduit or liquidtight flexible metal conduit is used as permitted and is to be relied on to complete a single equipment grounding path, it shall be installed with internal or ____ bonding jumpers in parallel with each conduit. Unless the conduit is six feet or less in length.

(a) factory installed (b) external (c) field soldered (d) field welded

62. The branch-circuit conductors supplying one or more units of a data processing system shall have an ampacity not less than _____ of the total connected load.

(a) 75% (b) 100% (c) 110% (d) 125%

63. Where the driving machine of an electric elevator or the hydraulic machine of a hydraulic elevator is located in a remote machine room or remote machinery space, a ______ means for disconnecting all ungrounded main power supply conductors shall be provided and be capable of being locked in the open position.

(a) multiple (b) duplicate (c) local (d) single

64. When running fire alarm cable exposed, the cable must be secured no less then every ____ inches when the cables are within 7 feet of the floor.

(a) 5 (b) 10 (c) 18 (d) 24

65. Selecting a primary ____ location to achieve the shortest practical primary protector grounding conductor will help limit potential differences between communications circuits and other metallic systems.

(a) limiter (b) protector (c) neutral (d) common

66. Which of the following is not an approved method of running Class 2 and Class 3 circuits in hoistways?

(a) In rigid nonmetallic conduit **(b) In rigid metal conduit**
(c) In flexible metal conduit **(d) In electrical metallic tubing**

67. Switches, circuit breakers, relays, contactors, fuses and current-breaking contacts for bells, horns, sirens, and other devices in which sparks or arcs may be produced shall be provided with enclosures identified for ____ location.

(a) Class I (b) Class II (c) Class III (d) Class IV

68. The grounding conductor for communication circuits shall be connected to the ____.

I. metallic power raceway II. service equipment enclosure III. building electrode system

(a) I only (b) II only (c) II or III only (d) I, II or III

69. When installing a non-power limited fire alarm circuits the overcurrent protection shall not exceed ___ AMP for 18 AWG conductors and ____AMP for 16 AWG conductors.

(a) 7 AMP, 10 AMP (b) 8 AMP, 10 AMP (c) 6 AMP, 10 AMP (d) 15 AMP, 20 AMP

70. Adjustable-speed motors controlled by means of field regulation, shall be so that they cannot be started under a weakend field unless the motor is ____ for such starting.

(a) equipped (b) designed (c) connected (d) controlled

2005

OPEN BOOK EXAM #3

70 QUESTIONS
TIME LIMIT - 2 HOURS

TIME SPENT ☐ **MINUTES**

SCORE ☐ **%**

MASTER OPEN BOOK EXAM #3 **Two Hour Time Limit**

1. A 10 volt signal circuit is operated by a pushbutton switch in a Class I, Division 1 hazardous location as defined by the NEC. As a minimum, to protect against ignition of flammable vapors, this circuit must be run from the switch enclosure to outside of the hazardous location ____.

(a) in threaded rigid metal conduit with no couplings or fittings in between
(b) in threaded steel intermediate metal conduit with no couplings or fittings in between
(c) in threaded steel intermediate metal conduit with a conduit seal placed within 18" of the enclosure
(d) None of the above since low voltage (10v) may be present without the need for explosion protection

2. Underground wiring for bulk storage plants shall be installed in ____.

(a) threaded rigid metal conduit
(b) threaded steel intermediate metal conduit
(c) rigid non-metallic conduit buried not less than 24"
(d) Any of the above

3. In a theater dressing room, each switch controlling a receptacle shall be provided with a pilot light to indicate ____.

(a) when the receptacle is in use
(b) when the dressing room is in use
(c) when the receptacle is energized
(d) the ampere rating of the receptacle

4. The color coding for single-phase isolated circuits within a hazardous anesthetizing location is ____.

(a) Isolated Conductor No.1 -- Black
Isolated Conductor No.2 -- Blue
(b) Isolated Conductor No.1 -- Black
Isolated Conductor No.2 -- Red
(c) Isolated Conductor No.1 -- Orange
Isolated Conductor No.2 -- Brown
(d) Isolated Conductor No.1 -- Yellow
Isolated Conductor No.2 -- Brown

5. A grounding electrode system consisting of a ground ring shall encircle a building and consist of at least 20 feet of bare copper conductor not smaller than ____.

(a) #2 (b) #4 (c) #6 (d) #8

6. A motor controller which is subject to periodic submersion in water requires a NEMA Type ____.

(a) 3 (b) 4X (c) 6 (d) 12

7. Any pipe or duct system foreign to the electrical installation shall not enter or pass through a transformer vault. Which of the following is not considered as foreign to the electrical installation?

(a) A roof drainage pipe.
(b) The building water main.
(c) An automatic sprinkler system.
(d) An enclosed air duct passing through the vault.

8. Capacitors containing more than 3 gallons of flammable liquid shall be ____.

(a) enclosed with a fence not less than 6' in height
(b) enclosed with a fence with 5' of fence fabric and three or more strands of barbed wire
(c) enclosed with a fence not less than 6' 6" in height
(d) enclosed with a fence with 6' of fence fabric and three or more strands of barbed wire

9. A 950 volt transformer has a circuit breaker installed on the primary side. The circuit breaker shall be set or rated at no more than ____ of the primary current rating of the transformer in a supervised location.

(a) 250% (b) 300% (c) 400% (d) 600%

10. Single-speed motors starting on "Y" connection and running on delta connection shall be marked with a code letter corresponding to the locked-rotor kVA per horsepower for the ____.

(a) Y connection
(b) delta connection
(c) highest running speed
(d) maximum torque percentage

11. Outlet boxes identified for such use shall be permitted to support listed ceiling fans that do not exceed ____ pounds in weight.

(a) 16 (b) 33 (c) 35 (d) 50

12. Recessed HID light fixtures operated by remote ballasts shall be thermally protected ____.

(a) at the point of supply
(b) at the fixture only
(c) at the ballast location only
(d) at the fixture and the ballast

13. Multiconductor portable cables over 600 volts nominal used to connect mobile equipment and machinery shall employ flexible stranding. The conductors shall be ____ copper or larger.

(a) #12 (b) #10 (c) #8 (d) #6

14. An AC-DC general-use snap switch may be used with INDUCTIVE loads at ____ of its ampere rating at the applied voltage.

(a) 30% (b) 50% (c) 75% (d) 80%

15. Switching devices consisting of combined temperature-actuated devices and manually controlled switches that serve both as the controllers and the disconnecting means shall comply with all the following conditions:

(a) Open all grounded conductors when manually placed in the "off" position.
(b) Be designed so that the circuit cannot be energized automatically if the device is manually placed in the "on" position.
(c) Be provided with a positive lockout in the "off" position.
(d) Be provided with a positive lockout in the "on" position.

16. DC motors operating from a rectifier bridge of the single-phase full-wave power supply, the conductors between the field wiring terminals of the rectifier and the motor shall have an ampacity of not less than ____of the motor full-load current rating.

(a) 120% (b) 125% (c) 150% (d) 190%

17. A part-winding connected motor, the selection of branch-circuit conductors on the line side of the controller shall be based on the motor full-load current. The selection of conductors between the controller and the motor shall be based on ____ of the motor full-load current.

(a) 50% (b) 58% (c) 100% (d) 125%

18. Audio equipment, storage batteries, transformers, transformer rectifiers and other AC & DC power supplies other than branch-circuit power supplies shall be installed and ___ in accordance with the requirements of this code for the voltage and power delivered.

(a) programmed (b) wired (c) removed (d) de-energized

19. For spraying operations confined to an enclosed spray booth, the Division 2 area extends ____ in all directions from any opening.

(a) 3'
(b) 10'
(c) 900'
(d) unlimited

20. The recreational vehicle back-in site electrical supply equipment shall be located on the _____ side of the parked vehicle; on a line that is 1.5 m to 2.1 m (5 ft to 7 ft) from the left edge of the stand. And shall be located at any point on this line from the rear of the stand to 4.5 m (15 ft) forward of the rear of the stand.

(a) left (b) front (c) rear (d) right

21. If the exhaust ventilation system is not interlocked with the spray application equipment, then the Division 2 location shall extend 3 m or ____ ft. horizontally and 900 mm (3 ft) vertically from the open face or open front of the booth or room.

(a) 3 (b) 8 (c) 10 (d) 12

22. Aircraft batteries shall not be _____ where installed in an aircraft located inside or partially inside a hangar.

(a) charged (b) removed (c) connected (d) de-energized

23. Portable lighting equipment in commercial garages shall be equipped with ____ and substantial guard attached to the lampholder or handle. All exterior surfaces that might come in contact with battery terminals, wiring terminals, or other objects shall be of nonconducting material or shall be effectively protected with insulation.

(a) handle (b) receptacle (c) AFCI protection (d) identified for Class III locations

24. Where shielded conductors or cables are used, shields shall be ______. Exception: Where a shield is part of an intrinsically safe circuit.

(a) connected (b) terminated (c) grounded (d) bonded

25. Pendant lighting fixtures shall be suspended by threaded rigid metal conduit stems, where stems are longer than 300 mm (12 in.). Permanent and effective bracing against lateral displacement shall be provided at a level not more than ____ above the lower end of the stem.

(a) 12 in. (b) 14 in. (c) 18 in. (d) 24 in.

26. Switches, circuit breakers, motor controllers, and fuses, including push buttons, relays, and similar devices that are intended to interrupt current during normal operation or that are installed where combustible dusts of an electrically conductive nature may be present, they shall be provided with identified dust ignition proof _____.

(a) labels (b) enclosure (c) sign (d) warnings

27. Fire alarm circuits shall be identified at the______ and _______ locations in a manner that will prevent unintentional interference with the signaling circuit during testing and servicing.

(a) terminal and junction **(b) conduit and j-box**
(c) beginning and end **(d) termination and end**

28. The roof, floors, and doorways of vaults containing conductors and equipment over 600 volts, nominal, shall be constructed of materials that have adequate structural strength for the conditions, with a minimum fire rating of __hrs.

(a) 2 (b) 3 (c) 4 (d) 6

29. A system in which heat is generated in a pipeline or vessel wall by causing current to flow through the pipeline or vessel wall by direct connection to an ac voltage source from a dual-winding transformer is called a ____ heating system.

(a) integrated (b) induction (c) impedance (d) direct

30. The ampacity of branch circuit conductors shall not be less than _____ percent of the designated potential load of utilization equipment that will be operated simultaneously.

(a) 50 (b) 75 (c) 100 (d) 125

31. Portable handlamps shall comply with which of the following conditions before they are permitted to be used.

(a) Shall not be wired with flexible cord.
(b) Metal shell, paper-lined lampholders shall be used.
(c) Handle of molded composition or other insulating material.
(d) Handlamps with a substantial guard and not grounded if metal.

32. Devices such as ________ shall not be used for the protection of conductors against overcurrent due to short circuits or ground faults, but the use of such devices shall be permitted to protect motor branch-circuits.

(a) fuses (b) thermal relays (c) breakers (d) a, b, and c

33. Each lighting and appliance branch-circuit panelboard shall be individually protected on the supply side by not more than _____ main circuit breakers or two sets of fuses having a combined rating not greater than that of the panel board.

(a) one (b) two (c) three (d) four

34. Of the following items listed below applies to what needs to be marked on attachment plugs and cord connectors.

(a) year mfg. (b) voltage (c) watts (d) frequency

35. Metal enclosures shall be protected both in-side and outside against ________.

(a) tampering (b) corrosion (c) wear (d) overfill

36. Circuit breakers used for over current protection of three-phase circuits over 600v shall have a minimum of ___ over current relay elements operated from the current transformer.

(a) one (b) two (c) three (d) four

37. For an angle pull, the distance between each cable or conductor entry inside the box and the opposite wall of the box shall not be less than _______ times the outside diameter, over sheath, of the largest cable or conductor. This distance shall be increased for additional entries by the amount of the sum of the outside diameters, over sheath, of all other cables or conductor entries through the same wall of the box, for over 600 volts.

(a) 4 (b) 24 (c) 30 (d) 36

38. The voltage developed between the portable or mobile equipment frame and ground by the flow of maximum ground-fault current shall not exceed ______ volts.

(a) 50 (b) 100 (c) 125 (d) 200

39. Bends in type TC cable without metal shielding, the minimum bending radius shall be _______ times the overall diameter for cables larger than 50 mm (2 in.) in diameter.

(a) six (b) five (c) four (d) three

40. Switches shall be marked with _______ and some times horsepower with the maximum rating for which they are designed.

(a) watts (b) current voltage (c) hertz (d) on position

41. Temperature controlled switching devices that do not have the ______ position shall not be required to open all ungrounded conductors and shall not be permitted to serve as the disconnecting means.

(a) off (b) auto (c) on (d) reverse

42. Where the ends of a conduit terminate in areas of widely differing temeratures, ____.

(a) conduit ends should be filled with an approved material
(b) rigid metallic conduit should be used
(c) electrical metallic conduit should be used
(d) flexible metallic conduit should be used

43. Disregarding exceptions, direct burial cable rated above ____ volts shall be shielded.

(a) 600 (b) 1000 (c) 1500 (d) 2000

44. Cables rated over 600v may be installed in the same cable tray with cables rated 600v or less ____.

(a) Where cables over 600v are Type MC
(b) where separated by a solid fixed barrier of a material compatible with the cable tray
(c) both (a) and (b)
(d) neither (a) nor (b)

45. Open conductors shall be separated from metal conduits, piping or other conducting materials ny no less than ____.

(a) 8" (b) 6" (c) 4" (d) 2"

46. Type UF cable shall be permitted ____.

(a) in a theater
(b) in commercial garages
(c) where embedded in poured concrete
(d) for interior wiring in wet or corrosive locations

47. According to the National Electrical Code, disregarding any exceptions, rigid nonmetallic conduit shall not be used ____.

(a) in cinder fill
(b) for underground work
(c) for the support of fixtures
(d) in walls, floors and ceilings

48. Where liquidtight flexible metal conduit is installed as a fixed raceway, it shall be secured at intervals not exceeding 4 1/2' and within ____ on each side of every outlet box or fitting.

(a) 6" (b) 8" (c) 10" (d) 12"

49. Outlet boxes mounted in combustible walls or ceilings must be mounted so they will be ____ the finished surface.

(a) set back not more than 1/8" from
(b) set back not more than 1/4" from
(c) set back not more than 1/2" from
(d) flush or project from

50. Power sources other than transformers used for Class 1 Power-Limited Circuits shall be protected by overcurrent devices rated at not more than ____ of the volt-ampere rating of the source divided by the rated voltage.

(a) 75% (b) 80% (c) 133% (d) 167%

51. Taps to brake coils of cranes shall be required to have separate overcurrent protection of ____.

(a) the sum of their nameplate current rating of all motors
(b) not less than one-third of the branch circuit
(c) in accordance with 430.72
(d) no separate overcurrent protection is required

52. When there is a power outage in a hospital and the automatic switching equipment is activated to emergency service, what is the required time delay setting for re-establishment of the normal service when normal power is restored?

(a) Not to exceeds 10 seconds
(b) 15 minutes
(c) 1 1/2 hours
(d) 2 hours

53. Bonding conductors for signs shall be not smaller than ____.

(a) #14 Al (b) #14 Cu (c) #12 Cu (d) #10 Cu

54. If a remote-control circuit for safety-control equipment would introduce a direct fire or life hazard if the equipment should fail, this is considered a Class ____ circuit by the Code.

(a) 1 (b) 2 (c) 3 (d) Not classified

55. All electric pool water heaters shall have the heating elements subdivided into loads not exceeding 48 amperes and protected at not more than ____ amperes.

(a) 48 (b) 55 (c) 60 (d) 75

56. All metal parts of a swimming pool structure shall be bonded together . Which of the following is not a Code requirement?

(a) The wall of a bolted or welded metal pool shall be permitted as a common bonding grid.
(b) A solid copper #8 copper conductor shall be permitted as a common bonding grid.
(c) The #8 bare copper bonding conductor shall be connected shall be connected to the equipment grounding conductor.
(d) The structural reinforcing steel of a concrete pool where the reinforcing rods are bonded together by the steel tie wires or the equivalent.

57. In a Class 1 circuit, overcurrent protection for #16 conductors shall not exceed ____ amperes.

(a) 5 (b) 7 (c) 10 (d) 15

58. Transfer equipment may be used to provide an alternate source of electrical power ____.

(a) in an emergency system
(b) in a gasoline dispensing system
(c) in a grounding system of a building
(d) where there exists combustible dust in explosive conditions

59. A complete system consisting of components such as pipelines, vessels, heating elements, heat transfer medium, thermal insulation, moisture barrier, nonheating leads, temperature controllers, safety signs, junction boxes, raceways, and fittings is called a _____ heating system.

(a) integrated (b) induction (c) impedance (d) direct

60. Adjustable Speed Drive Systems the disconnecting means shall be permitted to be in the incoming line to the conversion equipment and shall have a rating not less than ____ percent of the rated input current of the conversion unit.

(a) 115 (b) 125 (c) 150 (d) 175

61. Each 125-volt single-phase 15 and 20 ampere receptacle installed in pits, in hoistways, on elevator car tops, and in escalator and moving walk wellways shall be _____ circuit interruption type to protect personnel.

(a) REN (b) GFCI (c) FRN (d) NON 205

62. Circuit conductors supplying conversion equipment included as part of an adjustable-speed drive system shall have an ampacity not less than ____ of the rated input to the power conversion equipment.

(a) 120% (b) 125% (c) 150% (d) 190%

63. A controller shall not be required to open all conductors to the motor, unless the controller serves as the _____ for the motor as well.

(a) relay (b) contactor (c) controller (d) disconnecting means

64. Hoistways and cars shall have there traveling cables supported by which of the following methods.

(a) Support is not required is less than 100 feet.
(b) By looping the cables around supports for unsupported lengths less than 30 m (100 ft).
(c) By suspending from the supports by a means that automatically tightens around the cable when tension is increased for unsupported lengths up to 120 m (400 ft).
(d) Support is not required if less than 75 feet.

65. Ballast, transformers, and electronic power supplies shall be permitted to be located in attics and soffits, provided there is an access door and passage way at least ________ . As well as a suitable permanent walkway at least 300 mm (12 in.) wide, extending from the point of entry to each component.

(a) 3' x 3' (b) 2 1/2' x 3' (c) 3' x 2' (d) 2' x 3 1/2'

66. Hoistway pits must have at least ______ duplex receptacle provided to aid in service or clean up of the pit.

(a) one (b) two (c) three (d) zero

67. Where flexible connections are necessary to be used on cranes or hoist, flexible ______ conductors shall be used.

(a) solid (b) braided (c) stranded (d) multi

68. A minimum of 70 percent of all recreational vehicle sites with electrical supply shall each be equipped with a _____ ampere, 125-volt receptacle.

(a) 15 (b) 20 (c) 15 (d) 30

69. The only pipe or duct allowed to enter or pass through the transformer vault shall be _____.

(a) fire protection (b) alarm protection (c) ground detection (d) none allowed

70. Conductors and cables of intrinsically safe circuits not in raceways or cable trays shall be separated at least ______ and secured from conductors and cables of any nonintrinsically safe circuits.

(a) 24" (b) 12" (c) 6" (d) 2"

2005

OPEN BOOK EXAM #4

70 QUESTIONS
TIME LIMIT - 2 HOURS

TIME SPENT ☐ **MINUTES**

SCORE ☐ **%**

MASTER OPEN BOOK EXAM #4 Two Hour Time Limit

1. For over 600v, the motor branch circuit conductors shall have an ampacity not less than ___.

(a) 175% of the motor nameplate current
(b) 150% of the motor nameplate current
(c) 140% of the full load current from the appropriate table in the NEC
(d) the current at which the motor overload device is selected to trip

2. The ampacity of a single #12 fixture wire is ____ amps.

(a) 20 (b) 25 (c) 23 (d) 35

3. Embedded deicing units shall not exceed ____ watts per square foot of heating area.

(a) 120 (b) 15 (c) 16.5 (d) 33

4. A 20 amp branch circuit is installed to supply receptacles for stage set lighting. The receptacles connected to this circuit shall have a minimum rating of ___ amperes.

(a) 15 (b) 20 (c) 25 (d) 15 or 20

5. Resistors and reactors rated over 600v shall be isolated by ___.

I. elevation II. enclosure III. barrier tape

(a) I only (b) II only (c) I & II only (d) II & III only

6. Exposed snow melting heating elements shall be secured on the heated surface by a ____ means.

(a) rigid (b) approved (c) listed (d) stable

7. The minimum depth of oil over the power contacts in an oil immersion type switch for use in a Class I Division II location is ___ inches.

(a) 1 (b) 1 1/2 (c) 2 (d) 3

8. Each storage or instantaneous type water heater shall be equipped with a ___ to disconnect all ungrounded conductors and such means shall be installed to sense maximum water temperature and be either a trip free, manually reset type or have replaceable elements.

(a) thermostat (b) disconnect (c) relay (d) temperature limiting means

9. For a wye-start delta-run connected motor the conductors between the controller and a 3 hp, 3ø, 208v motor shall have an ampacity of at least ____ % of the motor F.L.C.

(a) 150 (b) 115 (c) 125 (d) 58

10. Each patient bed location shall be supplied by at least ___ branch circuits, equally supplied by the emergency system and the normal system.

(a) two (b) four (c) six (d) eight

11. The rated full load current for a DC motor, 7 1/2 hp, 500v would be ____ amps.

(a) 11 (b) 13 (c) 13.6 (d) 80

12. Where rigid nonmetallic conduit is used for the forming shell of an underwater speaker to a suitable junction box, a ___ conductor shall be installed in this conduit and properly terminated at the forming shell and the junction box for protection from the possible deteriorating effect of pool water.

(a) #12 copper insulated (b) #10 copper uninsulated
(c) #8 insulated (d) #14 insulated

13. Transformer ventilation shall be ____ to dispose of full load losses.

(a) sufficient (b) rated (c) adequate (d) derived

14. Where 3-phase power is supplied, a separately derived ____ system with 60 volts to ground installed for sensitive electronic equipment shall be configured as three separately derived 120-volt single-phase systems having a combined load of no more than six main disconnects.

(a) 3-phase delta (b) 3-phase wye (c) 6-phase delta (d) 6-phase wye

15. A rotary phase converter is a device having a rotary transformer and ____ panel (s) that can operate 3ø loads from a 1ø source.

(a) capacitor (b) secondary (c) primary (d) regulator

16. The cord exposed usable length shall be measured from the point of entrance to the park trailer or the face of the flanged surface inlet (motor base attachment plug) to the face of the attachment plug at the supply end. The maximum length shall not exceed ___ feet.

(a) 23 (b) 28 (c) 32 1/2 (d) 36 1/2

17. In an elevator machine room, at least ___ receptacle(s) shall be installed.

(a) one duplex (b) two duplex (c) one single (d) two receptacles, opposite walls

18. A capacitor's residual voltage shall be reduced to ____ volts or less within a 5 minute time period after being disconnected from the source.

(a) 100 (b) 135 (c) 150 d) 50

19. A phase converter disconnecting means shall be ____.

(a) switch rated in hp (b) circuit breaker
(c) molded case switch (d) all of these

20. The electric vehicle supply equipment cable shall have an overall length not exceeding ___ feet.

(a) 12 (b) 25 (c) 50 (d) 100

21. In Class II locations ____ dust may dehydrate or carbonize making them even more dangerous.

(a) plastic (b) coal (c) organic (d) metallic

22. In an office with relocatable partitions, individual partitions or groups of interconnected individual partitions shall not contain more than ___ 15 ampere, 125 volt receptacle outlets.

(a) 9 (b) 10 (c) 13 (d) unlimited

23. An electric mixer intended for traveling in and out of an open mixing tank shall be considered ____ utilization equipment.

(a) fixed (b) stationary (c) portable (d) mobile

24. Non-heating leads of snow melting equipment shall have a minimum length of not less than ____ inches within the junction box.

(a) 3 (b) 4 (c) 6 (d) 8

25. The power conductors in type NMS cable are manufactured in sizes ___.

(a) #14 - #6 (b) #14 - #4 (c) #14 - #2 (d) #12 - #2

26. Factory assembled nonmetallic underground conduit with conductors is not permitted ___.

(a) in cinder fill
(b) in exposed indoor locations
(c) encased or imbedded in concrete
(d) in underground locations subject to severe corrosive influences

27. A receptacle is permitted in a Class I Division II circuit if ____.

I. attachment plug is not depended on to interrupt current
II. current does not exceed 3 amps at 120 volts nominal
III. cord does not exceed 3 feet
IV. only necessary receptacles, provided with warning labels

(a) I and II only (b) I and IV only (c) I, II and III only (d) I, II, III and IV

28. At a gasoline dispensing, self service station which is unattended, the emergency disconnect for a circuit leading to dispensing equipment, must be located more than ___ feet from the dispensers.

(a) 3 (b) 5 (c) 10 (d) 20

29. For a motor an instantaneous trip circuit breaker shall be used any if ___and if part of a listed combination motor controller having coordinated motor overload and short circuit and ground fault protection.

(a) permitted by the local jurisdiction (b) available (c) adjustable (d) damping equipped

30. Receptacles on construction sites shall not be connected to the same ___ conductor of multiwire circuits that supply temporary lighting.

(a) grounded (b) ungrounded (c) grounding (d) none of these

31. Boxes and fittings shall be ____ in a Class III, Division I location.

(a) sealed (b) dustproof (c) dusttight (d) not allowed

32. All metal surfaces that are within 5 feet of the inside walls of a spa or hot tub and not separated from the spa or hot tub area by a permanent barrier shall be bonded together except ___.

I. drain fitting attached to metal piping
II. metal fittings attached to the spa or hot tub structure
III. towel bars

(a) I only (b) II only (c) III only (d) I, II and III

33. Fixed lighting fixtures for anesthetizing locations shall be at least ____ feet above the floor.

(a) 12 **(b) 10** **(c) 8** **(d) 5**

34. When installing a type FCC system under carpet squares, not more than ___ crossings of cable runs shall be permitted at any one point.

(a) 1 (b) 2 (c) 4 (d) 5

35. The temperature rating associated with the ampacity of a conductor shall be so selected and coordinated as to not exceed the ___ temperature rating of any connected termination, conductor, or device.

I. lowest II. highest III. ambient

(a) I only (b) II only (c) II & III only (d) I, II or III

36. The ampacity for branch circuit conductors supplying X-ray equipment marked (50 amps momentary) would require an ampacity of at least ____ amps.

(a) 25 **(b) 40** **(c) 50** **(d) 62.5**

37. For elevator systems that regenerate power back into the power source, which is unable to absorb the regenerative power under overhauling elevator load condition, ___.

(a) a transfer switch shall be installed to connect the regenerative power to the main power source
(b) a disconnect switch shall be labeled "caution regenerative power available"
(c) a means shall be provided to absorb this power
(d) the elevator shall be locked out immediately

38. Bare bulbs installed backstage that may come into contact with combustible material shall be guarded and provided with at least ____ inches of air space.

(a) 1 1/2 **(b) 2** **(c) 6** **(d) 12**

39. An equipotential plane shall be installed where required to mitigate ____ at electrical equipment.

(a) step and touch voltages **(b) stray currents**
(c) inductance **(d) capacitance**

40. A fixture wire used for controling X-ray equipment shall have a minimum conductor size of ____.

(a) #20 **(b) #18** **(c) #12** **(d) #8**

41. The walls and roofs of transformer vaults shall be constructed of materials that have adequate structural strength for the conditions with a minimum fire resistance of 3 hours unless protected by automatic sprinklers. A typical 3-hour construction is ___.

(a) reinforced concrete 6 inches thick (b) reinforced concrete 4 inches thick
(c) reinforced concrete 3 inches thick (d) studs and wall board construction

42. An assembly that has concealed parts from process of manufacturing, and cannot be inspected before being installed at a building site without disassembly, damage, or destruction, is a definition of _____.

(a) enclosed (b) guarded (c) closed construction (d) inaccessible

43. Taps to individual motors on an irrigation machine branch circuit supplying three 2 hp motors, 3ø, 460v, shall have a minimum size of at least ____ copper.

(a) #8 (b) #10 (c) #12 (d) #14

44. A/an ___ is a jack or terminal bus that serves as the collection point for redundant grounding of electric appliances serving a patient vicinity or for grounding other items in order to eliminate electromagnetic interference problems.

(a) hazard current (b) exposed conductive surface
(c) isolated power system (d) patient equipment grounding point

45. A/an ___ is a unitized segment of an industrial wiring system in which orderly shutdown is necessary to ensure safe operation.

(a) emergency standby electrical system (b) selective load pick-up electrical system
(c) critical branch electrical system (d) integrated electrical system

46. Where the maximum branch circuit short circuit and ground fault protective device ratings are shown in the ___ overload relay table for use with a motor controller or are otherwise marked on the equipment, they shall not be exceeded even if higher values are allowed.

(a) UL (b) manufacturer's (c) standard (d) Code

47. The maximum height for the coupling means to an indoor electric vehicle charging station is ____ feet.

(a) 4 (b) 4 1/2 (c) 8 (d) 10

48. On a delta connected 3ø, 4-wire system the phase arrangement for the busbar with the highest voltage to ground is ____ phase.

(a) neutral (b) C (c) A (d) B

49. Free standing office partitions are permitted to have outlets supplied by a single flexible cord as long as ____.

(a) extra hard usage type are used
(b) #12 or larger conductors are used with an insulating equipment grounding conductor
(c) the cord does not exceed 2' in length
(d) all of these

50. ___ shall be placed in all service entrance motor control centers to isolate service busbars and terminal from the remainder of the motor control center.

(a) Barriers (b) Shunt trips (c) In line fuses (d) HACR breaker

51. Roof mounted photovoltaic arrays located on dwellings shall be provided with ___ to reduce fire hazard.

(a) ground fault protection **(b) automatic extinguishing system**
(c) alarms **(d) smoke detectors**

52. A stage set lighting load of 50,000 va will be allowed a feeder demand factor of ____.

(a) 50% (b) 60% (c) 75% (d) 100%

53. If a ___ serves as the disconnecting means for a permanently connected motor- driven appliance of 1/8 horsepower or more, it shall be located within sight from the motor controller.

I. circuit breaker II. switch

(a) I only (b) II only (c) either I or II (d) neither I or II

54. Cablebus shall be securely supported at intervals not exceeding ___ feet.

(a) 6 (b) 10 (c) 12 (d) 18

55. The allowable ampacity of a listed extra-hard usage #14 cord with a temperature rating of 90°C (194°F) is ___ amperes used for border lights and based on an ambient temperature of 30°C, 86°F and the cable is not in direct contact with equipment containing heat-producing elements.

(a) 14 (b) 25 (c) 28 (d) 20

56. ___ electrical equipment enclosures shall be permitted where a natural draft ventilation system prevents the accumulation of gases at cell lines.

(a) Rear accessible (b) Isolating type (c) General purpose (d) Flush mounted

57. Auxiliary gutters shall be permitted to supplement wiring spaces at meter centers, distribution centers, switchboards, and similar points of wiring systems and may enclose ___.

I. switches II. overcurrent devices III. conductors IV. busbars

(a) I & II only (b) I & III only (c) I & III only (d) III & IV only

58. For over 600v busways having sections located both inside and outside of buildings shall have a ___ at the building wall.

(a) vapor seal (b) fire barrier (c) condulet (d) ventilated enclosure

59. For signaling circuits not exceeding 24 volts, the current required shall not exceed ___ ampere where protected by an overcurrent device or an inherently limited power source.

(a) 1 (b) 4 (c) 6 (d) 10

60. Where coaxial cable is directly buried for a community antenna television system, the cable shall be separated at least ___ inches from conductors of any light or power or Class 1 circuit.

(a) 2 (b) 6 (c) 12 (d) 24

61. Where a transformer vault is constructed with other stories below it, the floor shall have a minimum fire resistance of ___ hours.

(a) 4 (b) 3 (c) 2 (d) 1 1/2

62. Fire alarm systems include___.

I. sprinkler water flow II. alarm notification III. guard's tour

(a) I only (b) II only (c) I and II only (d) I, II, and III

63. A park trailer is intended to be for ___.

I. seasonal use II. a temporary office III. a permanent dwelling

(a) I only (b) II only (c) III only (d) I, II, or III

64. ___ are typically used as enclosures for splicing or terminating telephone cables.

I. Splice cases II. Terminal boxes III. Octagonal boxes

(a) I only (b) II and III only (c) I and II only (d) none of these

65. The operating grounding conductor for amateur transmitting stations shall not be less than #___ copper.

(a) 18 (b) 16 (c) 14 (d) 10

66. The terminal block for the termination of flat cable assemblies shall have distinctive and durable marking for color or word coding. The grounded conductor section shall have a white marking or other suitable designation. The next adjacent section of the terminal block shall have a ___ marking.

(a) black (b) blue (c) red (d) green

67. An electrical contractor has received the plans and specifications for a skating rink. The National Electrical Code rules governing this occupancy are in Article ___.

(a) 518 (b) 520 (b) 710 (d) 820

68. In a swimming pool, electric equipment that depends on submersion for safe operation shall be protected against overheating by a/an ___.

(a) toggle switch (b) thermal overload (c) in line fuse (d) low water cut off

69. All branch circuit that supply 125-volt, single-phase, 15 and 20 ampere outlets installed in dwelling unit bedrooms shall be protected by an arc-fault circuit interrupter listed to provide protection of the _____.

(a) receptacles (b) lighting outlets (c) entire branch circuit (d) the entire feeder

70. Bridge wire contact conductors for cranes shall be kept at least 2 1/2 inches apart, and where the span exceeds ___ feet, insulating saddles shall be placed at intervals not exceeding ___ feet.

(a) 50 - 80 (b) 60 - 80 (c) 80 - 60 (d) 80 - 50

2005

OPEN BOOK EXAM #5

70 QUESTIONS
TIME LIMIT - 2 HOURS

TIME SPENT ☐ MINUTES

SCORE ☐ %

MASTER OPEN BOOK EXAM #5 Two Hour Time Limit

1. In a solar photovoltaic system, where the grounding connection is not opened automatically as part of the ground-fault protection system, a switch or circuit breaker used as a disconnecting means shall ____.

(a) not have a pole in the grounded conductor
(b) be required to open the grounded conductor
(c) be permitted to open the grounded conductor
(d) have an overcurrent device in the grounded conductor

2. Which of the following is not an approved method of running Class 2 and Class 3 circuits in hoistways?

(a) In rigid nonmetallic conduit **(b) In rigid metal conduit**
(c) In flexible metal conduit **(d) In electrical metallic tubing**

3. Single conductor portable supply cable sets to portable stage switchboards shall be permitted to have the grounded and grounding conductors identified by marking at least the first ____ from both ends with appropriate colored marking.

(a) 3" (b) 6" (c) 9" (d) 12"

4. Manhole covers shall be over ____ pounds or otherwise designed to require the use of tools to open.

(a) 50 (b) 75 (c) 80 (d) 100

5. The branch circuit supplying the elevator car lights and accessories shall ____.

(a) be a dedicated circuit for all the cars
(b) be a separate branch circuit for each car
(c) be connected to the elevator feeder circuit
(d) be any normal lighting circuit in the building

6. Power-limited fire protective signaling circuit conductors are to be installed within an enclosure containing electric light and power conductors. Which of the following best describes Code requirements for these conductors?

(a) They must be installed within their own compartment.
(b) They must never enter enclosures with other conductors.
(c) They must be separated at least 0.50 inches from other conductors.
(d) They must be separated at least 0.25 inches from other conductors.

7. The conductors which extend from the connection point of the serving utility at the top of a building to the service disconnecting means are called _____.

(a) the service lateral **(b) the service drop**
(c) the service conductors **(d) the service point**

8. A branch-circuit panelboard is to be installed in a dwelling unit. The most convenient location is in a room which contains a toilet and shower stall. Which of the following best describes Code requirements concerning the location of this panelboard?

(a) It is not permitted in this room.
(b) It may be installed in this room.
(c) The location of the panelboard is not covered by Code.
(d) It may be installed in this room, but it must be weatherproof.

9. The lower (basement) level of a dwelling unit is completely finished off into habitable rooms except for an 8' by 8' area which contains the heater and water heater. A duplex receptacle is to be installed in this area. Which of the following best describes Code requirements for installation of this receptacle?

(a) It is not permitted to be installed in this area.
(b) It is required but does not require GFCI protection.
(c) The Code does not cover 8' by 8' areas in basements.
(d) It is required to be installed and must have GFCI protection.

10. For a hermetic refrigerant motor-compressor, the rated-load current marked on the nameplate shall be used in determining ____.

(a) ampacity of disconnecting means **(b) branch-circuit conductors**
(c) separate motor overload protection **(d) all of these**

11. In general, the overcurrent protection for the single-phase supply conductors to a phase converter shall not exceed 125% of the ____.

(a) phase converter nameplate three-phase full-load amperes
(b) total sum of three-phase loads served by the phase converter
(c) phase converter nameplate single-phase input full-load amperes
(d) none of the above

12. Outdoor self-supporting radio and television receiving antennas, such as vertical rods, dishes, etc., shall be located well away from overhead conductors of electric light and power circuits of over ____ volts to ground.

(a) 30 (b) 120 (c) 150 (d) 300

13. Conductive noncurrent carrying parts of electrical equipment are bonded to the grounded circuit conductor of the system for which of the following purposes?

(a) To protect against lightning damage.
(b) To make the equipment operate at 120v.
(c) To facilitate overcurrent device operation during short-circuit.
(d) To facilitate overcurrent device operation during ground-fault.

14. For small stationary motors, of 2 hp or less and 300v or less, the disconnecting means can be ____.

(a) the branch-circuit breaker **(b) a stop/start station**
(c) a general use snap-switch **(d) none of these**

15. A commercial building has an opening door in the wall on the second floor through which materials are moved into and out of a storage area located on the second floor. An overhead 120v branch circuit is to be installed from the second floor area of the building to a pole on which a flood-light will be mounted. Which of the following best describes Code requirements for the installation of this branch-circuit?

(a) It is not permitted to be installed overhead.
(b) It must be at least 3' from the side of the door.
(c) It must be at least 3' above the bottom of the door.
(d) It must not obstruct entrance to the material handling door and be 3' from the door.

16. When AC cable is installed in thermal insulation it shall have conductors rated at ____.

(a) 60° C (b) 75° C (c) 90° C (d) 105° C

17. A 120v, 20 amp duplex receptacle is installed in an elevator pit to supply power for a sump pump. Which of the following best describes Code requirements for this installation?

(a) The receptacle must be a single outlet, no GFCI required.
(b) The receptacle must be on a dedicated circuit.
(c) A receptacle is not permitted in an elevator pit.
(d) The receptacle must have GFCI protection for personnel.

18. A metal faceplate is installed on a wall switch in the stairway to an attic space. The wiring method is nonmetallic sheathed cable with an equipment grounding conductor. There are no grounded surfaces within 8' of this switch. Which of the following best describes Code requirements for this installation?

(a) The metal plate should be replaced with a plastic plate.
(b) The metal plate does not require grounding.
(c) A metal plate is not permitted to be installed.
(d) The metal plate shall be grounded.

19. The nearest readily accessible location for the main 150 amp service disconnecting means for a dwelling unit is in the bathroom. The required working clearances can be maintained in this room and the branch-circuit overcurrent devices will be located in another room through a feeder. Which of the following best describes Code requirements for this installation of this disconnecting means?

(a) It is not permitted in this room.
(b) It may be installed in this room.
(c) The location of disconnecting means is not covered by Code.
(d) It may be installed in this room, but must be a weatherproof type.

20. A two-wire 125v duplex wall receptacle in the living room of a dwelling unit must be replaced due to damage. There is no grounding means in the metal box which contains the receptacle. Which of the following best describes Code requirements for replacement of this device?

(a) An ungrounded three-wire device must be installed.
(b) A two-wire device must be installed as the replacement.
(c) A three-wire grounded device must be installed as the replacement.
(d) An ungrounded three-wire device may be used with a GFCI breaker.

21. Branch-circuit conductors within 3 inches of a ballast in the ballast compartment shall have an insulation temperature rating of not less than ____ degrees Celsius.

(a) 90 (b) 105 (c) 167 (d) 194

22. When fixture sections are paired, with a ballast supplying lamps in both, it shall be permissible to use 3/8" flexible metal conduit in lengths not exceeding ____ feet to interconnect the paired units.

(a) 6 (b) 10 (c) 15 (d) 25

23. A duplex receptacle is installed on a 4" square raised metal cover and mounted on a 4" square metal box. If the grounding terminal of the receptacle is connected to the grounded metal box, which of the following requirements would also apply to the installation of the duplex receptacle?

(a) It must be secured by at least two screws.
(b) It is not permitted to be installed in raised covers.
(c) It must be secured directly to the 4" square box.
(d) The single 6-32 screw holding it in the cover must be tight.

24. Which of the following best describes Code requirements for the installation of spas and hot tubs?

(a) All installations require GFCI protection.
(b) Only indoor installations require GFCI protection.
(c) Only cord and plug connected installations require GFCI protection.
(d) Only outdoor installations require GFCI protection.

25. Which of the following is an acceptable means of preventing tension from being transmitted to the joints or terminals of a flexible cord?

(a) Knotting the cord **(b) Winding with tape**
(c) Special fittings **(d) All of these**

26. A metal pole supporting a light fixture is installed in a parking lot. The pole is 18 feet high but does not have a 2 x 4 inch handhole in the bottom. Which of the following best describes Code requirements for this installation?

(a) The pole must have a hinged base.
(b) Metal poles are not permitted to support light fixtures.
(c) A pole manufactured without a handhole, is okay to install.
(d) A pole over 8 feet high requires a 2 x 4 inch handhole in the base.

27. A health care facility which is used on a 24 hour basis for housing 4 or more persons who are incapable of self-preservation because of age, physical limitation due to accident or illness, or mental limitations would be defined as a ____.

(a) hospital (b) nursing home (c) limited care facility (d) psychiatric hospital

28. A piece of liquidtight flexible nonmetallic conduit 6 feet long is used for a raceway to contain service entrance conductors for a 208v wye service. Which of the following best describes Code requirements for installation of this liquidtight flexible nonmetallic conduit?

(a) It is approved without a bonding jumper.
(b) It is not permitted for service raceways.
(c) It is approved for this use only on dwelling units.
(d) An equipment bonding jumper can be installed inside or outside the conduit.

29. An existing wall receptacle in the men's room of an office building must be replaced due to damage. Which of the following best describes Code requirements for this replacement?

(a) A receptacle is not permitted in office bathrooms.
(b) A standard grounding-type duplex receptacle may be used.
(c) A GFCI protected receptacle must be used as the replacement.
(d) None of the above.

30. Electrically operated waste disposers intended for dwelling use and which are cord and plug connected shall have a cord which is not less than ____ inches long nor more than 36" long.

(a) 12 (b) 16 (c) 18 (d) 24

31. A 120v circuit is installed to provide power to the pump for an LP-Gas filling station located at a gasoline dispensing service station. The LP-Gas will be dispensed into portable tanks as used in barbeque grills. Which of the following best describes Code requirements for the disconnecting means for this circuit?

(a) The Code does not cover LP-Gas dispensing stations.
(b) A 240v disconnect must be used instead of 120v.
(c) LP-Gas cannot be dispensed at a gasoline service station.
(d) The disconnect must also disconnect the grounded conductor.

32. Enclosures that are not over 100 cubic inches in size and which have two conduits supported within three feet on either side of the enclosure and the enclosure does not contain devices or support fixtures shall not be required to have the enclosure supported if the conduits are ____.

(a) rigid nonmetallic conduits
(b) threaded into hubs identified for the purpose
(c) installed with locknuts inside and outside enclosure
(d) shoulders of fittings outside and locknuts inside the box

33. A #10 copper equipment grounding conductor can be used for branch circuits up to ____ amps.

(a) 30 (b) 40 (c) 50 (d) 60

34. A bank of capacitors, rated at 60 amperes, is connected in a circuit. What is the minimum size 75° Celsius copper conductors allowed for making this connection?

(a) #3 (b) #4 (c) #8 (d) #6

35. 100 feet of rigid nonmetallic conduit is run between two cabinets. If its thermal expansion is less than ____ inches, an expansion joint is not required.

(a) .25 (b) .28 (c) .30 (d) .35

36. Temporary wiring for trade show display booths in exhibition halls shall be ____.

(a) GFCI protected
(b) permitted to be laid on floors
(c) installed in rigid conduit
(d) supported 10 feet off the floor

37. Any switchboard which is not fully enclosed requires a space of ____ feet or more between it and any combustible ceiling.

(a) 2.0 (b) 3.0 (c) 3.5 (d) 4.0

38. A #8 copper grounding electrode conductor for a 100 ampere service is installed to a metal water pipe. Which of the following best describes Code requirements for this installation?

(a) It is only permitted to connect to a ground rod.
(b) It must be enclosed in a raceway or cable armor.
(c) It is not permitted to be used as a grounding electrode conductor.
(d) None of the above.

39. "The conductor used to connect the non-current carrying metal parts of equipment, raceways, and other enclosures to the system grounded conductor and/or to the grounding-electrode conductor at the service equipment or at the source of separately derived system", is called the ____.

(a) bonding jumper **(b) grounded conductor**
(c) equipment-bonding jumper **(d) equipment-grounding conductor**

40. When supporting romex parallel to framing members, such as joists, rafters, or studs, how close to the nearest edge can it be installed?

(a) 1 inch (b) 1 1/4 inches (c) 1 1/2 inches (d) 2 inches

41. Taps from a 50 ampere branch circuit must be a minimum ____ copper.

(a) #14 (b) #12 (c) #10 (d) #8

42. An 800 amp high-impedance grounded neutral 480v electrical system is installed in an industrial plant. The grounding impedance will limit the current flow to 8 amperes. The minimum size of the neutral conductor from the neutral connection to the grounding impedance shall not be less than ____.

(a) #8 (b) #6 (c) #1/0 (d) #2/0

43. A portable outdoor electric sign that is readily accessible shall have ground-fault protection for personnel. This protection shall be ____.

(a) factory installed as part of the plug or cord
(b) a ground-fault circuit-interrupter circuit breaker
(c) a ground-fault circuit-interrupter type receptacle
(d) any type GFCI protection

44. Rigid nonmetallic conduit shall be supported so that ____.

(a) the conduit will not move
(b) the conduit can move from expansion
(c) the conduit can move from contraction
(d) the conduit can move from expansion and contraction

45. A feeder is to be installed in a health care facility which has a large 120/208v wye service. The service equipment main has ground-fault protection installed. Which of the following best decribes Code requirements for this new feeder?

(a) The feeder must have ground-fault protection.
(b) The feeder must have fully selective ground-fault protection.
(c) A new feeder is not permitted to be installed in an old building.
(d) Ground-fault protection is only required on 480v wye systems.

46. A receptacle is installed outdoors on the porch of a dwelling unit. The receptacle is located such that it is near the open railing of the porch, but is at a height of 7 feet above grade level. Which of the following best describes Code requirements for installation of this receptacle?

(a) It is not permitted on the porch.
(b) It is not permitted close to an open railing.
(c) It must have GFCI protection as it is outside.
(d) It does not require GFCI protection even though it is outside.

47. A #8 bare copper conductor has a circular mil area of _____.

(a) 10,380 (b) 16,510 (c) 26,240 (d) 41,740

48. Splices and taps in surface nonmetallic raceways shall not fill the raceway at the point of the splice or tap to more than ____.

(a) 20% (b) 50% (c) 75% (d) 80%

49. Where the number of current-carrying conductors in a raceway equals 10, the allowable ampacities shall be reduced to ____ of the values listed in Tables 310.16 through 310.19.

(a) 45% (b) 50% (c) 60% (d) 80%

50. What is the recommended maximum percent voltage drop on both feeders and branch circuits to the farthest outlet?

(a) 2.5% (b) 3% (c) 3.5% (d) 5%

51. A 125v, 20 amp, single-outlet type, wall receptacle is to be installed in the bathroom of a drug store. Which of the following best describes Code requirements for the installation of this receptacle?

(a) It must have GFCI protection.
(b) It must be a duplex, not a single outlet.
(c) It is not permitted in a drug store bathroom.
(d) It does not require GFCI protection because it is a single outlet.

52. Unbroken lengths of surface nonmetallic raceways may pass through which of the following?

(a) Dry walls (b) Damp walls (c) Damp floors (d) Wet locations

53. A 3 foot piece of liquidtight flexible metal conduit is installed in a service entrance raceway which contains 4 pieces of #500 kcmil copper for a 400 amp, 480v wye service. A bonding jumper will be installed on the outside of this liquidtight flexible metal conduit. What is the minimum size of this bonding jumper?

(a) #1 copper (b) #2 copper (c) #3 copper (d) 1/0 copper

54. Mandatory rules of the NEC are characterized by the use of the word "shall". Explanatory materials are in the form of ____.

(a) NEMA (b) FPN (c) exceptions (d) UL

55. What size main bonding jumper is required to bond a separately derived system supplied from a 225 kva, 480v delta, 120/208v wye transformer where the secondary conductors are parallel #350 kcmil THHN supplying a 600 amp distribution panel?

(a) #1 copper (b) #2 copper (c) #1/0 copper (d) #2/0 copper

56. A coaxial cable for a community antenna television (CATV) is installed in a ditch with a 120v type UF cable to a building. The Code requires that these two cables have a minimum separation of ____ .

(a) 6" (b) 12" (c) 18" (d) no separation required

57. What is the hp rating of a single-phase 200v motor with a current rating of 11.5 amps?

(a) 1.5 hp (b) 3 hp (c) 5 hp (d) 7.5 hp

58. An electrical substation operating at over 600v is constructed at the rear of a building. The fence enclosing this equipment must be a minimum of ____ high.

(a) 6 feet (b) 7 feet (c) 8 feet (d) 10 feet

59. A duplex wall receptacle is to be installed in the garage of a dwelling unit. The receptacle will be located between a refrigerator and a freezer so that each are plugged into the duplex receptacle. Which of the following best describes Code requirements installation of this receptacle?

(a) It is not permitted in a garage.
(b) It must have GFCI protection because it is in a garage.
(c) GFCI protection is not required as it serves two appliances.
(d) One single-type receptacle must be installed for each appliance.

60. An equipment-bonding jumper is permitted to be installed on the outside of a raceway or enclosure, but it must not exceed ____ in length.

(a) 3 feet (b) 6 feet (c) 10 feet (d) 12 feet

61. If two motors are supplied by the same branch-circuit, the ampacity of the circuit conductors must be equal to the sum of the full-load current ratings of both motors plus ____ of the largest motor.

(a) 25% (b) 75% (c) 110% (d) 125%

62. A motor disconnecting means shall be within sight from the ____.

(a) motor and have a lockable stop/start button
(b) controller, but must have a lock-out at the motor
(c) motor location only if the disconnect or stop/start is lockable
(d) controller and disconnect all ungrounded conductors into it

63. The minimum size copper conductor permitted in metal-clad cable is ____.

(a) #18 (b) #16 (c) #14 (d) #12

64. The grounding electrode conductor of a grounded 400 amp AC system, when connected to a made electrode and which is the sole connection to the grounding electrode, is permitted to be a ____.

(a) #6 copper (b) #4 copper (c) #2 copper (d) #1/0 copper

65. A circuit in which any arc or thermal effect produced, under intended operating conditions of the equipment due to opening, shorting, or grounding, etc., is not capable of igniting the flammable gas, vapor, or dust-air mixture is called a ____.

(a) overload device **(b) nonincendive circuit**
(c) multi-wire circuit **(d) separately derived system**

66. A piece of electronic equipment is required to be connected with an isolated equipment grounding conductor. The design is such that the metal raceway containing the supply conductors is to be isolated from the equipment enclosure, but the raceway contains an insulated copper equipment grounding conductor. Which of the following best describes Code requirements for this installation?

(a) The raceway shall not be isolated from the equipment.
(b) The isolation fitting may be located at the panelboard.
(c) The isolation fitting may be anywhere in the conduit run.
(d) The isolation fitting must be located at the equipment enclosure.

67. Lighting track is permitted to be installed in which of the following areas?

(a) in wet or damp locations **(b) where concealed**
(c) extended through walls **(d) six feet above the finished floor**

68. The maximum fixture weight that can be supported by an outlet box is ____ pounds.

(a) 35 (b) 40 (c) 45 (d) 50

69. In general, Class 2 and Class 3, remote control and signaling circuit wiring shall not be secured to ____.

(a) raceways (b) metal studs (c) wood studs (d) other low voltage conductors

70. The grounded conductor of a two-wire branch circuit shall be identified by a ____.

(a) continuous white color **(b) white or green markings**
(c) continuous green color **(d) continuous white or natural gray color**

2005

OPEN BOOK EXAM #6

70 QUESTIONS
TIME LIMIT - 2 HOURS

TIME SPENT ☐ **MINUTES**

SCORE ☐ **%**

MASTER OPEN BOOK EXAM #6 Two Hour Time Limit

1. Access doors or detachable access panels shall be employed for internal access to heating equipment. Access doors to internal compartments containing equipment employing voltages from ____ ac or dc shall be capable of being locked closed or shall be interlocked to prevent the supply circuit from being energized while the induction heating door is open.

(a) 60 v to 120 v (b) 120 v to 240 v (c) 240v to 600v (d) 150v to 1000v

2. HDPE conduit smaller than trade size ____ shall not be used.

(a) 3/8" (b) 1/2" (c) 5/8" (d) 3/4"

3. Coaxial cables for a fire alarm system shall be permitted to have copper covered steel center conductors with a minimum of 30 percent copper and shall be listed as Type ___.

I. FPL II. FPLR III. FPLS IV. FPLP

(a) I and II only (b) II or III only (c) I, II, or IV only (d) I, II, III, or IV

4. Conductors installed in conduit exposed to direct sunlight in close proximity to rooftops have been shown, under certain conditions, to experience a temperature rise of ____ above ambient temperature on which the ampacity is based.

(a) 30°C (b) 86°F (c) 17°C (d) 90°C

5. In fixed electrostatic spray zones, signs shall be conspicuously posted to ___.

I. restrict access to qualified personnel only
II. identify emergency exit routes
III. designate the zone as dangerous with regard to fire and accident
IV. identify the grounding requirements for all electrically conductive objects in the spray area

(a) I only (b) I and II only (c) I, III, and IV only (d) I, II, III, and IV

6. In a park trailer, lighting switches shall be rated not less than ___ amperes 120/125 volts and in no case less than the connected load.

(a) 10 (b) 15 (c) 20 (d) 25

7. A Class 1 power-limited circuit shall be supplied from a source having a rated output of not more than 30 volts and ___ volt amperes.

(a) 1000 (b) 1200 (c) 1500 (d) 2000

8. Transformers insulated with listed less flammable liquids having a fire point of not less than 300° C shall be permitted to be installed in Type I or Type II buildings if the transformer is rated ___ volts or less.

(a) 600 (b) 7200 (c) 35,000 (d) 69,000

9. Emergency system(s) overcurrent devices shall be selectively ____ with all supply side overcurrent protective devices.

(a) interchangeable (b) GFCI protected (c) for manual operation (d) coordinated

10. Insulation of conductors intended for use in the secondary circuit of electric welders shall be ___.

I. flame retardant II. silicone III. TBS IV. TFE

(a) I only (b) I & II only (c) II & III only (d) I, & IV only

11. When installing X-ray equipment connected to 120 volt branch circuit, a grounding type attachment plug cap and receptacle of proper rating shall be permitted to serve as a disconnecting means if the branch circuit is rated ___ amperes or less.

(a) 15 (b) 20 (c) 30 (d) 50

12. A subsystem of the emergency system consisting of feeders and branch circuits, meeting the requirements of article 700 and intended to provide adequate power needs to ensure safety to patients and personnel, and which are automatically connected to alternate power sources during interruption of the normal power source is the ___.

(a) essential system (b) critical branch (c) life safety branch (d) line isolation monitor

13. A dry cell battery shall be considered an inherently limited ___ power source, provided the voltage is 30 volts or less and the capacity is equal to or less than that available from series connected No. 6 carbon zinc cells.

(a) Class 1 (b) Class 2 (c) Class 3 (d) none of these

14. Where stage smoke ventilators are released by an electrical device, the circuit operating the device shall be ___ and shall be controlled by at least two externally operable switches.

(a) a pressure switch (b) a limit switch (c) normally closed (d) normally open

15. A fire alarm circuit that is powered by an output voltage of not more than 600 volts is defined as a ___ fire alarm circuit.

(a) NPLFA (b) PLFA (c) FPLP (d) FPLF

16. Premise power distribution systems jointly controlled by a signaling between the energy controlling equipment and utilization equipment is a ___ system.

(a) remote control, signaling, and power limited circuit
(b) fire alarm
(c) communication circuit
(d) closed loop and programmed power distribution

17. To interrupt currents over 1200 amperes at 250 volts, or over 600 amperes at 251 to 600 volts, a ___ of special design listed for such purpose shall be used.

I. switch II. circuit breaker III. bypass isolation switch

(a) I & II only (b) II & III only (c) II only (d) I, II, or III

18. Open runs of braid-covered insulated conductors used for over 600 volts shall have a flame retardant saturant applied to the braid covering after installation unless this protectant is factory installed. This treated braid covering shall be stripped back a safe distance at conductor terminals, a distance of not less than ___ for each kilovolt of the conductor to ground voltage of the circuit.

(a) 1" (b) 2" (c) 3" (d) 6"

19. An AC adjustable voltage motor that does not have the maximum operating current on the nameplate, shall have the ampacity of the branch circuit conductors based on ___ of the values given in Tables 430.149 and 430.150.

(a) 80% (b) 100% (c) 125% (d) 150%

20. Individual office partitions of the freestanding type or groups of individual partitions that are electrically connected, mechanically contiguous, and do not exceed ___ feet when assembled, shall be permitted to be connected by a single flexible cord and plug.

(a) 10 (b) 15 (c) 24 (d) 30

21. Coaxial cable shall be permitted to deliver low energy power to equipment directly associated with radio frequency distribution system if the voltage is not over ___ volts and if the current supply is from a transformer or other device having energy limiting characteristics.

(a) 24 (b) 30 (c) 60 (d) 70.7

22. Any interruption of the circuit, even circuits as low as ___ volts, either by any switch, or loose or defective connections anywhere in the circuit, may produce a spark sufficient to ignite flammable anesthetic agents.

(a) 10 (b) 12 (c) 24 (d) 30

23. Portable switchboards and portable power distribution equipment in an exhibition hall shall ___.

I. be supplied only from listed power outlets of sufficient voltage and ampere rating
II. be protected by overcurrent devices accessible to the general public
III. be provided with provisions for connection of an equipment grounding conductor

(a) I only (b) I and II only (c) I and III only (d) I, II, and III

24. For the purpose of this Code and unless otherwise indicated, the term "mobile home" includes ___.

I. park trailers II. travel trailers III. recreational vehicle IV. manufactured homes

(a) I only (b) II and III only (c) IV only (d) I and IV only

25. Branch circuits for heating and air conditioning equipment located on the elevator car shall not have a circuit voltage in excess of ___ volts.

(a) 120 (b) 240 (c) 480 (d) 600

26. Supply circuit conductors to a fire pump shall ___.

I. be physically routed outside the building
II. be routed through the building where installed under or enclosed within 1" of concrete
III. be routed through the building where installed under or enclosed within 2" of concrete

(a) I only (b) II only (c) I or II only (d) I or III only

27. Wall mounted remote control stations for remote control switches operating at ___ or less shall be permitted to be installed in any anesthetizing location.

(a) 12 volts (b) 24 volts (c) 30 volts (d) 60 volts

28. A ___ arranged to be locked in the open position shall be provided in the leads from the runway contact conductors or other power supply on all cranes and monorail hoists.

I. mushroom head push button II. motor circuit switch III. circuit breaker

(a) I or II only (b) II only (c) II or III only (d) I, II, or III

29. Conductors that supply one or more welders shall be protected by an overcurrent device rated or set at not more than ___ of the conductor rating.

(a) 80% (b) 125% (c) 150% (d) 300%

30. For small motor compressors not having the locked rotor current marked on the nameplate, or for small motors not covered by the FLC tables, the locked rotor current shall be assumed to be ___ times the rated load current.

(a) 1 1/2 (b) 1 3/4 (c) 2 1/2 (d) 6

31. A bonding jumper not smaller than ___ copper or equivalent shall be connected between the antenna systems grounding electrode and the power grounding electrode system at the building or structure served where separate electrodes are used.

(a) #12 (b) #10 (c) #8 (d) #6

32. Size #18 or #16 fixture wires, and flexible cords shall be permitted for the control and operating circuits of X-ray and auxiliary equipment where protected by not larger than ___ ampere overcurrent devices.

(a) 6 (b) 15 (c) 20 (d) 30

33. All manual switches for controlling emergency lighting in a theater or motion picture theater shall be located in ___.

I. projection booth II. on the side of the stage or platform III. in the lobby

(a) I only (b) II only (c) III only (d) II or III only

34. For a photovoltaic power source, one conductor of a 2 wire system rated over ___ volts and a neutral conductor of a 3 wire system shall be solidly grounded.

(a) 50 (b) 60 (c) 90 (d) 100

35. A fuel cell system that supplies power independently of an electrical production and distribution network is a ______.

(a) interactive system (b) stand-alone system (c) one-fer system (d) independent network

36. Sealed battery cells shall be equipped with a ___ or the battery cell shall be designed to prevent scatter of cell parts in event of a cell explosion.

I. flame arrester II. pressure release vent

(a) I only (b) II only (c) both I and II (d) none of these

37. A transformer vault ventilated by natural circulation of air to an outdoor area, shall have a ventilating opening not less than 1 sq. ft. for any capacity under ___ kVA.

(a) 25 (b) 50 (c) 75 (d) 150

38. When a cord and plug connection is provided for connecting lighting equipment for use with relocatable wired office partitions, the cord length shall not be longer than ___ feet.

(a) 3 (b) 6 (c) 9 (d) 12

39. Duty on wheelchair lift and stairway chair lift driving machine motors shall be rated as ___.

(a) continuous (b) intermittent (c) varying (d) short-time

40. A building or part thereof used on a 24 hour basis for the housing of four or more persons who are incapable of self preservation because of age, physical limitation due to accident or illness, or mental limitations, such as mental retardation/developmental disability, mental illness or chemical dependency is a/an ___.

(a) limited care facility (b) nursing home
(c) psychiatric hospital (d) ambulatory health care center

41. Where single conductor cables in sizes #1/0 through 500 kcmil are installed, continuously covered for more than 6 feet with unventilated covers, the ampacities shall not exceed ___percent of the allowable ampacities in Tables 310.17 and 310.19.

(a) 50 (b) 60 (c) 65 (d) 80

42. Flat cable assemblies may be installed ___.

I. for small power loads outdoors, not subject to physical damage
II. as tap devices for lighting and small appliances
III. for small power loads in hoistways

(a) I only (b) II only (c) I and III only (d) I, II, and III

43. In panelboards, where close proximity of bare metal parts such as busbars does not cause excessive heating, parts of the same polarity at switches, enclosed fuses, etc., shall be permitted to ___.

(a) be placed a minimum of 1/2" apart
(b) be placed a minimum of 3/4" apart
(c) be placed a minimum of 1 1/4" apart
(d) be placed as close together as convenience in handling will allow

44. All high voltage parts, including X-ray tubes, shall be mounted within ___ enclosures.

(a) metallic (b) nonmetallic (c) grounded (d) ventilated

45. It shall be permissible to use 3/8" flexible metal conduit in lengths not exceeding ___ feet for interconnection between paired fixture sections operating at 1000 volts or less.

(a) 6 (b) 12 (c) 18 (d) 25

46. ___ is/are permitted to be installed for deck area heating at a swimming pool

I. Radiant heaters that are installed more than 5 feet from the inside walls of the pool
II. Unit heaters that are rigidly mounted and are installed more than 5 feet from the inside walls of the pool
III. Radiant heating cables that are embedded in the concrete deck a minimum of 2 inches in depth

(a) I only (b) I and II only (c) II and III only (d) I, II, or III

47. On ac circuits, ___ shall be permitted to disconnect a motor rated 2 horsepower or less and 300 volts or less having a full load current rating not more than 80 percent of the ampere rating of the switch.

(a) general use ac/dc snap switches **(b) the branch circuit overcurrent device**
(c) general use snap switch suitable only for ac **(d) none of these**

48. In theaters and similar locations, ___ is/are permitted wiring method(s).

I. EMT II. NM cable III. MI cable IV. nonmetallic raceways encased in 2" of concrete

(a) I and II only (b) II and III only (c) III and IV only (d) I, III, and IV only

49. Cabinets and cutout boxes that contain devices or apparatus connected within the cabinet or box to more than ___ conductors, including those of branch circuits, meter loops, feeder circuits, power circuits, and similar circuits, but not including the supply circuit or continuation thereof, shall have back wiring spaces or one or more side wiring spaces, side gutters, or wiring compartments.

(a) 8 (b) 10 (c) 21 (d) 30

50. A minimum of ___ percent of all recreational vehicle sites with electrical supply shall be equipped with a 30 ampere 125 volt receptacle.

(a) 5 (b) 25 (c) 50 (d) 70

51. Liquidtight flexible nonmetallic conduit and rigid nonmetallic conduit shall be permitted to enclose the leads to the motor provided ___.

I. the required equipment grounding conductor is connected to the motor
II. the required equipment grounding conductor is connected to the junction box to the motor
III. the leads are solid
IV. the leads are stranded and shall not be larger than #10

(a) I and II only (b) II and III only (c) I, II, and IV only (d) I, II, and III only

52. A wireway or auxiliary gutter installed in a theater shall not exceed ___.

I. 20 percent of the interior cross sectional area of the gutter or wireway
II. 40 percent of the interior cross sectional area of the gutter or wireway
III. not more than 30 current carrying conductors at any cross section of the gutter or wireway

(a) I only (b) II only (c) I and III only (d) I, II, and III

53. Unattended self service gasoline stations shall have emergency controls to shut off all power to dispensing equipment. These controls shall be located more than ___ feet but less than ___ feet from the dispensers.

(a) 20 - 100 (b) 20 - 50 (c) 10 - 100 (d) 10 - 50

54. When installing remote control signaling circuits, ___ shall not be considered safety control equipment.

I. household heating and air conditioning equipment
II. water temperature regulating devices
III. room thermostats

(a) I only (b) II only (c) III only (d) I, II, and III

55. A lighting fixture installed in or on the wall of a storable pool shall ___.

I. have a transformer with a primary rating not over 150 volts
II. have an impact resistant polymeric lens, fixture body, and transformer enclosure
III. have a fixture lamp that operates at 120 volts or less

(a) I only (b) II only (c) I and II only (d) II and III only

56. In closed loop distribution systems, listed hybrid cable consisting of power, communications, and signaling conductors shall be permitted under a common jacket. The signaling conductors shall not be smaller than # ___ copper.

(a) 16 (b) 18 (c) 22 (d) 24

57. The protective grounding conductor for transmitting stations shall be as large as the lead-in, but not smaller than #___ copper.

(a) 16 (b) 18 (c) 12 (d) 10

58. Where used at a point on a circuit, the TVSS shall be connected to ____ conductor.

(a) each grounded (b) each ungrounded (c) the grounding (d) the electrode

59. Secondary circuits of current and potential instrument transformers shall be grounded where the primary windings are connected to circuits of ___ volts or more to ground and, where on switchboards, shall be grounded irrespective of voltage.

(a) 30 (b) 100 (c) 250 (d) 300

60. The demand factor to be applied in sizing the feeder that supplies power to 7 elevators with a 50 percent duty cycle is ___ percent.

(a) 125 (b) .82 (c) .79 (d) .77

61. Remote crane or hoist controls that may introduce hazardous electrical conditions into the cell line working zone shall employ one or more of the following systems.

I. Isolated and ungrounded control circuit in accordance with section 668-21 (a).
II. Nonconductive rope operator.
III. Radio.

(a) I only (b) II only (c) III only (d) I, II, or III

62. In health care facilities, where ground fault protection is provided for operation of the service disconnecting means or feeder disconnecting means, an additional step of ground fault protection shall be provided in the next level of feeder disconnecting means downstream toward the load. The additional levels of ground fault protection shall not be installed ___.

I. on electrical systems that are not solidly grounded wye systems with greater than 150 volts to ground.
II. on the load side of an essential electrical system transfer switch.
III. between the on site generating unit and the essential electrical system transfer switch.

(a) I only (b) II only (c) II and III only (d) I, II, and III

63. A room air conditioner rated ___ shall not be cord and plug connected.

(a) over 240 volts (b) over 250 volts (c) over 3 hp (d) over 5 hp

64. The grounding electrode to which the portable or mobile equipment system neutral impedance is connected shall be isolated from and separated in the ground by at least ___ feet from any other system or equipment grounding electrode.

(a) 6 (b) 10 (c) 12 (d) 20

65. When installing open wiring on insulators in a dry location, conductors shall be permitted to be separately enclosed in flexible nonmetallic tubing. The tubing shall be in continuous length not exceeding ___ feet and secured to the surface by straps at intervals not exceeding 4 1/2 feet.

(a) 6 (b) 10 (c) 15 (d) 50

66. Communication wires and cables shall be separated at least ___ from conductors of any electric light or power circuits, Class 1, or non power-limited fire alarm circuits, or medium power network powered broadband communications circuits.

(a) 0" (b) 2" (c) 4" (d) 6"

67. The rated ampacity of conductors and overcurrent devices of a feeder for two or more branch circuits supplying X-ray units shall not be less than 100 percent of the momentary demand rating of the two largest X-ray apparatus plus ___ percent of the momentary ratings of other X-ray apparatus.

(a) 20 (b) 50 (c) 80 (d) 90

68. The equivalent continuous current rating for the selection of branch circuit conductors and branch circuit devices for a center pivot irrigation machine shall be equal to 125 percent of the motor nameplate full load current rating of the largest motor plus ___ of the sum of the motor nameplate full load current ratings of all remaining motors.

(a) 60% (b) 75% (c) 80% (d) 100%

69. The supply to a portable switchboard shall be by means of listed extra-hard usage cords or cables. Where connectors are used in the supply conductors, there shall be a maximum number of ___ interconnections (mated connector pairs) where the total length from supply to switchboard does not exceed 100 feet.

(a) 2 (b) 3 (c) 4 (d) 6

70. Indoor electric vehicle charging stations shall be provided with mechanical ventilation sized to provide the minimum ventilation in cubic feet per minute, for each parking space equipped to charge an electric vehicle. A 15 amp, 120 volt single phase charging space will require ___ CFM.

(a) 24 (b) 37 (c) 49 (d) 64

2005

OPEN BOOK EXAM #7

70 QUESTIONS
TIME LIMIT - 2 HOURS

TIME SPENT ______ **MINUTES**

SCORE ______ **%**

MASTER OPEN BOOK EXAM #7 Two Hour Time Limit

1. Exposed live parts within porcelain fixtures shall be suitably recessed and so located as to make it improbable that wires will come in contact with them. There shall be a spacing of at least ____ between live parts and the mounting plane of the fixture.

(a) 1/4" (b) 1/8" (c) 1/2" (d) 3/4"

2. All conduit referred to herein shall be threaded with a/an ____ standard conduit cutting die that provides 3/4" taper per foot.

(a) rigid (b) NPT (c) metal (d) American

3. Outlets for specific appliances such as laundry equipment, shall be within ____ feet of the appliance.

(a) 4 (b) 6 (c) 8 (d) 10

4. An enclosure designed either for surface or flush mounting and provided with a frame, mat, or trim in which a swinging door or doors are or may be hung is a ____.

(a) panelboard (b) switchboard (c) wireway (d) cabinet

5. Grounding of a metal raceway used to protect Romex is required if the raceway is ____ feet or over within reach of ground or grounded metal.

(a) 6 (b) 8 (c) 10 (d) 25

6. To prevent the entrance of moisture, service-entrance conductors shall be connected to the service-drop conductors ____.

I. below the level of the termination of the service-entrance cable sheath
II. below the level of the service head

(a) I only (b) II only (c) both I and II (d) neither I nor II

7. The area of square inches for a #6 bare conductor is ____.

(a) 0.017 (b) 0.027 (c) 0.207 (d) 0.184

8. Which of the following is **not** a standard size fuse?

(a) 110 amp (b) 125 amp (c) 75 amp (d) 250 amp

9. Which of the following is NOT considered an electric vehicle by the Code?

(a) industrial fork lift
(b) vans
(c) busses
(d) trucks

10. The N.E.C. covers ____.

I. gas welders II. DC rectifier arc welders III. motor-generator arc welders IV. resistance welders

(a) I and IV only (b) I, II and III only (c) II, III and IV only (d) I, II, III and IV

11. Type FCC cable wiring system is designed for installation under ____.

(a) tile (b) carpet (c) carpet squares (d) concrete

12. Service cables mounted in contact with a building shall be supported at intervals not exceeding ____ feet.

(a) 10 (b) 6 (c) 2 1/2 (d) 4 1/2

13. Multispeed motors shall be marked with the code letter designating the locked-rotor kva horsepower for the highest speed at which the motor ____.

(a) can be stalled **(b) can be started**
(c) needs to be rated **(d) can run safely**

14. Soft-drawn or medium-drawn copper, lead in conductors for receiving antenna systems shall be permitted where the maximum span between points of support is less than ____ feet.

(a) 35 (b) 30 (c) 20 (d) 10

15. Non-heating leads of heating cables operating in 240 volt systems, shall have a ____ color.

(a) red (b) blue (c) yellow (d) brown

16. Wading pools are those that are constructed on or above the ground and are capable of holding water to a maximum depth of ____.

(a) 18" (b) 30" (c) 42" (d) 4'

17. Branch circuit conductors within 3" of a ballast, within the ballast compartment shall be recognized for use at temperatures not lower than 90 degrees C, such as insulation types ____.

I. THHN II. THW III. TW IV. FEP

(a) I only (b) I and IV only (c) I, II and IV (d) I, II, III and IV

18. Amusement rides shall be maintained not less than ____ feet in any direction from overhead conductors operating at 600 volts or less, except for the supply conductors to the ride.

(a) 8 (b) 10 (c) 12 (d) 15

19. Flexible cord 50' in length shall be considered as protected by a 20 amp branch circuit overcurrent device if the cord is ____.

(a) not less than 6' in length
(b) #20 or larger
(c) #18 or larger
(d) #16 or larger

20. Infrared lamps for industrial heating appliances shall have overcurrent protection not exceeding ____ amps.

(a) 30 (b) 40 (c) 50 (d) 60

21. The temperature limitation of MI cable is based on the ____.

(a) ambient temperature
(b) conductor insulation
(c) insulating materials used in the end seal
(d) none of these

22. Overcurrent devices shall not be located in the vicinity of easily ignitible material such as in ____.

(a) bedrooms (b) clothes closets (c) kitchens (d) garages

23. A new building will have two service heads serviced by one service drop. What is the maximum distance apart that the Code permits these heads to be located?

(a) 36"
(b) 48"
(c) 6 feet
(d) there is no maximum distance as long as the service entrance conductors are long enough to reach the service drop conductors

24. Liquidtight flexible metal conduit is shipped in what sizes minimum and maximum?

(a) 1/2" to 4" (b) 1/2" to 6" (c)3/4" to 5" (d) 1/2" to 2"

25. The rated primary current of an electric resistance welder is ____.

I. 2.8 times the primary current divided by the duty cycle
II. determined by the percentage of time during which the welder is loaded
III. the rated kva multiplied by 1000 and divided by the rated primary voltage, using values given on the nameplate
IV. the current drawn from the supply circuit during each welder operation at the particular heat tap and control setting used

(a) I and III only (b) II and IV only (c) I, II and III (d) I, II, III and IV

26. Renewable type contacts shall be provided on all knife switches rated 600 volts designed for use in breaking current over ____ amperes.

(a) 50 (b) 100 (c) 150 (d) 200

27. Voltage shall not exceed 600 volts between conductors on branch circuits supplying only ballasts for electric-discharge lamps in tunnels with a height of not less than ____ feet.

(a) 12 (b) 15 (c) 18 (d) 22

28. 15 and 20 ampere receptacles located in pediatric areas shall be ____.

(a) tamper resistant (b) isolated (c) GFI (d) specification grade

29. A nipple contains six #8 THW copper current-carrying conductors. The ampacity of each conductor would be ____amperes.

(a) 24 (b) 50 (c) 35 (d) 40

30. Type MTW insulation would be used for ____.

(a) switchboards only (b) machine tool wiring
(c) feeders only (d) fixtures

31. Which of the following would **not** be approved in all Class II locations?

(a) flexible connections (b) threaded bosses
(c) dust-tight boxes (d) EMT

32. Emergency lighting, emergency power or both, in a building or group of buildings will be available within the time required for the application, but not to exceed one of the following:

(a) 5 seconds (b) 10 seconds (c) 30 seconds (d) 60 seconds

33. Aluminum grounding conductors shall not be used where ____.

I. within 18" of earth
II. subject to corrosive conditions
III. in direct contact with masonry

(a) I only (b) II only (c) III only (d) I, II and III

34. The heating of conductors depends on ____ values which, with generator field control, are reflected by the nameplate current rating of the motor-generator set driving motor rather than by the rating of the elevator motor, which represents actual but short-time and intermittent F.L.C. values.

(a) root-mean-square current (b) voltage
(c) ambient temperature (d) ozone

35. Each vented cell shall be equipped with a ____ designed to prevent destruction of the cell.

(a) gas arrestor (b) electrolyte (c) flame arrestor (d) hydrometer

36.Only composition or metal-sheathed, porcelain, keyless lampholders equipped with suitable means to guard lamps from physical damage and from film and film scrap shall be used at ____ tables.

I. cutting II. viewing III. patching

(a) I only (b) II only (c) III only (d) I, II and III

37. Parking garages used for parking or storage and where no repair work is done, open flame, welding, or the use of volatile flammable liquids are ____.

(a) Class I (b) Class II (c) Class III (d) not classified

38. The maximum weight of a light fixture that may be mounted on the screw shell of a brass socket is ____ pounds.

(a) 2 (b) 6 (c) 8 (d) 10

39. The basic photovoltaic device which generates electricity when exposed to light is a ____.

(a) battery (b) diode (c) solar cell (d) mandrel

40. The National Electrical Code scope includes ____.

(a) conductors and equipment on certain public or private premises
(b) conductors on all buildings or premises where electrical power is employed
(c) conductors and equipment on all buildings, premises and conveyances where electrical power is employed
(d) conductors and equipment wherever and whenever electricity is employed

41. Bare conductors on a 30 volt lighting system shall not be installed less than ___ feet above the finished floor, unless listed for a lower height.

(a) 7 (b) 8 (c) 9 (d) 10

42. For a closet that permits access to both sides of a hanging rod, the storage space shall include the volume below the highest rod extending ____ on either side of the rod on a plane horizontal to the floor extending the entire length of the rod.

(a) 6" (b) 8" (c) 12" (d) 15"

43. Open conductors on insulators must be covered when they are within ____ feet of a building.

(a) 10 (b) 12 (c) 15 (d) 25

44. Flexible cords shall be connected to devices and to fittings that tension will not be transmitted to joints or terminal screws. This shall be accomplished by ___.

(a) a knot in the cord **(b) winding with tape**
(c) a special fitting **(d) all of these**

45. Open conductors run individually as service drops shall be ____.

I. insulated II. bare III. covered

(a) I only (b) II only (c) III only (d) either I or III

46. In patient care areas of a hospital, the equipment grounding terminal bars of the normal and essential electrical system panelboards shall be bonded together with an insulated continuous copper conductor not smaller than ____.

(a) #8 (b) #6 (c) #10 (d) #12

47. The radius of a 2" conduit containing conductors without lead sheath shall not be less than ____ inches.

(a) 12 (b) 31 (c) 21 (d) 36

48. All heating cables, the splice between the heating cable and nonheating leads, and ____ minimum of the nonheating lead at the splice shall be embedded in plaster or dry board in the same manner as the heating cable.

(a) 3" (b) 6" (c) 8" (d) 12"

49. FCC cable connections shall use connectors identified for their use, such that ____ against dampness and liquid spillage are provided.

I. sealing II. insulation III. electrical continuity

(a) I only (b) II only (c) III only (d) I, II and III

50.The circuit supplying an autotransformer-type dimmer installed in theaters and similar places shall not exceed how many volts between conductors?

(a) 480 volts (b) 277 volts (c) 240 volts (d) 150 volts

51. For swimming pool water heaters rated at more than ___ amperes that have specific instructions regarding bonding and grounding, only those parts designated to be bonded shall be bonded, and only those parts designated to be grounded shall be grounded.

(a) 50 (b) 40 (c) 30 (d) 20

52. All fixed wiring above Class I locations in a repair garage shall be in ____.

I. flexible nonmetallic conduit II. rigid nonmetallic conduit III. TC cable

(a) I only (b) II only (c) I and II only (d) I, II and III

53. It shall be permissible to extend ____ of surface metal raceways through dry walls, dry partitions, and dry floors.

(a) 2 feet (b) 3 feet (c) 6 feet (d) unbroken lengths

54. A two pole circuit breaker that may be used for protecting a 3 phase corner grounded delta circuit shall be marked ___.

(a) 1ø 120/240v (b) 1ø—3ø (c) 1ø/2ø/3ø (d) 480Y/277v

55. Compliance with the provisions of the Code will result in ____.

(a) good electrical service **(b) an efficient system**
(c) freedom from hazard **(d) all of these**

56. In general, busways shall be supported at intervals of ____ feet.

(a) 3 (b) 5 (c) 6 (d) 10

57. All bonded parts of a pool shall be connected to a common bonding grid with a ____ conductor.

I. #8 stranded II. #6 solid aluminum III. #8 solid copper

(a) I only (b) II only (c) III only (d) I or III

58. The Code requires all conductors carrying AC current installed in metal raceways to be grouped together because ____.

(a) it's cheaper (b) it's easier to test (c) it's easier to maintain (d) of inductive current

59. Flexible cords to portable electrically heated appliances rated at more than ____ watts shall be approved for heating cords.

(a) 50 (b) 100 (c) 300 (d) 500

60. Electrical equipment provided with ventilating openings shall be installed so that walls or other obstructions do not prevent the ____ of air through the equipment.

(a) movement (b) release (c) free circulation (d) none of these

61. The size of the industrial control panel supply conductor shall have an ampacity not less than 125% of the full-load current rating of all resistance heating loads plus 125% of the full-load current rating of the highest rated motor plus the sum of the full-load current ratings of all other connected motors and apparatus based on their ____ that may be in operation at the same time.

(a) continuous load (b) varying duty (c) periodic duty (d) duty cycle

62. What is the maximum length of an unprotected feeder tap conductor?

(a) 10 feet (b) 25 feet (c) 50 feet (d) 100 feet

63. Electric ranges and clothes dryers used in mobile homes shall be ____.

(a) installed with the equipment enclosure ungrounded
(b) installed in the same manner as a dwelling unit
(c) installed with a grounding conductor run in parallel with the neutral
(d) installed with the grounded circuit conductor insulated from the grounding conductor and equipment enclosures

64. Cord-and-plug connected vending machines not incorporating integral GFCI protection shall be ____.

(a) installed on an insulating mat **(b) connected to a 15 amp breaker only**
(c) located 6' from any metallic object **(d) connected to a GFCI protected outlet**

65. Non-insulated busbars will have a minimum space of ____ inches between the bottom of enclosure and busbar.

(a) 6 (b) 8 (c) 10 (d)12

66. Wet-niche lighting fixtures shall be connected to an equipment grounding conductor not smaller than ____.

(a) #10 (b) #6 (c) #8 (d) #12

67. The minimum insulation level for neutral conductors of solidly grounded systems shall be ____ volts.

(a) 600 (b) 1000 (c) 1500 (d) 2100

68. The maximum operating temperatures of rubber-covered, type RFH-1 heat resistant fixture wire is ____ degrees F.

(a) 124 (b) 167 (c) 195 (d) 224

69. Essential electrical systems for hospitals shall be comprised of two separate systems. These two systems shall be the ____.

I. equipment system II. low voltage system III. emergency system

(a) I and II only (b) II and III only (c) I and III only (d) none of these

70. Instrument pilot lights and potential current transformers shall be protected by O.C.P. of ____ amps or less.

(a) 15 (b) 20 (c) 30 (d) 50

2005

OPEN BOOK EXAM #8

70 QUESTIONS
TIME LIMIT - 2 HOURS

TIME SPENT [ ] MINUTES

SCORE [ ] %

MASTER OPEN BOOK EXAM #8 — Two Hour Time Limit

1. Fixtures shall be securely fastened to ceiling framing member by mechanical means such as ____.

I. rivets II. screws III. bolts

(a) II only (b) III only (c) II and III only (d) I, II and III

2. Insulated conductors used in wet locations shall be ____.

(a) MTW (b) asbestos (c) THHN (d) varnished cambric

3. In dwelling units, a multiwire branch circuit supplying a split-wired receptacle is required to ____.

(a) have color-coded conductors to aid in identification
(b) carry an individual neutral for each ungrounded conductor
(c) have a provision at the source to simultaneously disconnect all ungrounded conductors
(d) have a minimum of two neutrals

4. The word transformer is intended to mean a ____ transformer, single or polyphase, identified by a single nameplate, unless otherwise indicated.

(a) group (b) two (c) individual (d) step-down

5. Unused openings in boxes, raceways, and other enclosures shall be ____.

(a) closed with a device listed for such service with the equipment
(b) effectively closed
(c) closed to afford protection equivalent to the equipment wall
(d) open

6. Spacing between conduits, tubing, or raceways shall be ____.

(a) 1/4" (b) insulated (c) 1" (d) maintained

7. Feeders to floating dwellings must be enclosed within ____ conduit in order to withstand the forces exerted by waves and tides.

(a) rigid metal (b) rigid PVC (c) liquidtight flexible (d) EMT

8. Running open wiring on insulators, MI or MC cable, messenger-supported wiring, conductors in raceway, and other approved means on the outdoor building surfaces is permitted for circuits operating at a maximum of ____ volts nominal.

(a) 600 (b) 750 (c) 1000 (d) 4160

9. Where a single AC conductor carrying current passes through metal with magnetic properties, the inductive effect shall be minimized by ____.

I. cutting slots in the metal between the individual holes through which individual conductors pass
II. passing all the conductors in the circuit through an insulating wall sufficiently large for all the conductors of the circuit

(a) I only (b) II only (c) both I and II (d) neither I nor II

10. For nondwelling units, it is permitted to use a ____ demand factor for that portion of a receptacle load that exceeds 10 kva.

(a) 70% (b) 80% (c) 50% (d) 40%

11. Wiring over and under navigable water must be approved by the ____.

(a) corps of engineers **(b) U.S. Coast Guard**
(c) authority having jurisdiction **(d) Department U.S. Navy**

12. The phase current in a grounding autotransformer is ____ the neutral current.

(a) twice (b) 1/2 (c) 1/3 (d) the same as

13. In a closed-loop and programmed power distribution system, the outlets shall ____.

(a) energize only when the load allows the completion of a loop
(b) be energized only when plugged-in equipment is identified
(c) be constantly energized, as is done in existing 120/240v power systems
(d) none of these

14. Electrical ducts shall include any of the electrical conduits recognized in Chapter 3 as suitable for use ____.

(a) over 600v (b) as bus bars (c) underground (d) exposed

15. All disconnect means required by the Code, and each service, feeder, and branch circuit at the point where it originates shall be legibly marked ____.

(a) with a sign "Danger of Electrocution" **(b) "Disconnect"**
(c) and provided with a lockout means **(d) to indicate its purpose**

16. For cables that have elliptical cross section, the cross-sectional area calculation shall be based on using ____ of the ellipse as a circle diameter.

(a) half (b) the radius (c) the major diameter (d) the circumference

17. The alternate source for emergency systems ____ be required to have ground-fault protection of equipment with automatic disconnecting means.

I. shall II. shall not

(a) I only (b) II only (c) either I or II (d) neither I nor II

18. Where two or more single-phase ranges are supplied by a 3-phase, 4-wire feeder, the total load shall be computed on the basis of ____ maximum number connected between any two phases.

(a) twice the (b) three times the (c) half the (d) none of these

19. Type ____ is a single or multiconductor solid dielectric insulated cable rated 2001 volts or higher.

(a) MI (b) NM (c) MC (d) MV

20. The secondary circuits of wound-rotor alternating current motors, including ____ shall be permitted to be protected against overload by the motor-overload device.

(a) resistors (b) controllers (c) conductors (d) all of these

21. Where secondary ties are used, an overcurrent device rated or set at not more than ____ percent of the rated secondary current of the transformers shall be provided in the secondary connections of each transformer.

(a) 100 (b) 150 (c) 250 (d) 300

22. Conductors supplying outlets for arc or Xenon projectors of the professional type shall not be smaller than ____.

(a) #12 (b) #10 (c) #8 (d) #6

23. Not more than ___ of high-voltage cable shall be run in a metal raceway or tubing from a transformer/power supply to other parts of the sign.

(a) 10' (b) 12' (c) 15' (d) 20'

24. Receptacles rated ____ amperes or less directly connected to aluminum conductors shall be marked CO/ALR.

(a) 20 (b) 25 (c) 30 (d) 50

25. Each commercial building and each commercial occupancy accessible to pedestrians shall be provided at an accessible location outside the entrance, with at least one ____ for sign or outline lighting use.

(a) outlet (b) duplex (c) GFCI (d) none required

26. Transformers and electronic power supplies shall have secondary current ratings not more than ____ milliamperes.

(a) 300 (b) 350 (c) 400 (d) 600

27. Electric-discharge lighting fixtures having exposed ____ shall be so installed that these parts will not be in contact with combustible material.

I. live parts II. ballasts or transformers III. auxiliary equipment

(a) I only (b) II only (c) III only (d) I, II and III

28. A/an ____ circuit is a circuit in which any spark or thermal effect is incapable of causing ignition of a mixture of flammable or combustible material in air under prescribed test conditions.

(a) low voltage (b) intrinsically safe (c) hazard-proof (d) explosive-proof

29. Only wiring methods recognized as ____ are included in the Code.

(a) approved (b) suitable (c) listed (d) identified

30. Where space is too limited to provide minimum clearances, such as at meters, panelboards, outlets, and switch points, concealed knob and tube wiring shall be enclosed in ____ which shall be continuous in length between the last support and the enclosure or terminal point.

(a) rigid metal conduit (b) EMT (c) IMC (d) flexible nonmetallic tubing

31. The N.E.C. covers ____.

I. installations of electrical conductors and equipment within or on public and private buildings or other structures, including mobile homes, recreational vehicles, and floating dwelling units; and other premises, such as yards, carnivals, parking and other lots, and industrial substations
II. installations of conductors that connect to the supply of electricity
III. installations of optic fiber cable

(a) I only (b) II only (c) III only (d) I, II and III

32. Circuits for ____ shall **not** be connected to any system containing trolly wires with a ground return.

(a) kitchens and laundry rooms **(b) carhouses and power houses**
(c) railway stations **(d) lighting and power**

33. When the kind of motor is single-phase AC or DC, and its supply system is 2-wire, single-phase AC or DC, one conductor grounded, the minimum number and the location of overload units, such as trip coils, relays or thermal cutouts shall be ____.

(a) two, one per phase in hot conductors **(b) one in the ungrounded conductor**
(c) one in each conductor **(d) two, one in each phase**

34. A box or conduit body shall not be required for each splice, junction, switch, pull, termination, or outlet points in wiring methods with removable covers such as ____.

I. surface raceways II. wireways III. cable trays

(a) I only (b) II only (c) I and II only (d) I, II and III

35. In land areas not subject to tidal fluctuation, the electrical datum plane is a horizontal plane ____ above the highest water level for the area occurring under normal circumstances.

(a) 2' (b) 3' (c) 4' (d) 8'

36. No ____ other than those specified as required for emergency use, shall be supplied by emergency lighting circuits.

I. appliances II. lamps III. fittings

(a) I only (b) I and II only (c) II and III only (d) I, II and III

37. The approximate area square inch of a #2 THW compact aluminum conductor is ____.

(a) .1473 (b) .1194 (c) .1182 (d) .1017

38. An appliance (not motor driven) is rated 1200 watts at 120 volts, with no marked nameplate, the branch circuit overcurrent device shall not exceed ____ amps.

(a) 15 (b) 20 (c) 30 (d) 40

39. A storage battery of suitable rating and capacity to supply and maintain at not less than ____ of system voltage the total load of the circuits supplying legally required standby power for a period of at least 1 1/2 hours.

(a) 100% (b) 75% (c) 50% (d) 87 1/2%

40. What is the maximum time of delay permitted for the GFI to operate where the ground-fault current is 4000 amperes?

(a) 1/2 second (b) 1 second (c) 3 seconds (d) 100 mili-seconds

41. Which of the following methods is **not** approved for conductor supports?

(a) deflecting of cables in junction boxes
(b) insertion of boxes
(c) clamping devices
(d) loop connectors

42. A separate branch circuit shall supply the ____ receptacles, auxiliary power source, and ventilation on each elevator car.

(a) motor (b) car lights (c) emergency phone (d) emergency exit

43. An underground service installed in PVC and having a 3" concrete envelope shall be buried a minimum of ____ inches.

(a) 6 (b) 12 (c) 18 (d) 24

44. The ground fault protection system shall be tested when it is ____.

(a) installed (b) energized for the first time (c) inspected (d) manufactured

45. Where a motor is operating and live parts of the motor controller have over 150 volts to ground and might be exposed to repairmen, what must be done for its safe maintenance?

(a) two technicians must be present when servicing the motor
(b) insulating mats shall be provided
(c) a sign saying "Danger" must be installed near the motor
(d) power disconnected when working on it

46. A junction box used in a system rated 1000 volts shall have a marking on the box of ____.

(a) Caution (b) Danger (c) Do Not Open (d) Danger High-Voltage Keep Out

47. Where single conductors #1/0 through 4/0 are installed in a ladder or ventilated trough cable tray they shall be installed in no more than ___.

I. a depth of 4" II. a depth of 6" III. a single layer

(a) I only (b) II only (c) III only (d) I or II only

48. According to the N.E.C., high-voltage service-entrance conductors are protected by a circuit breaker if it has ____ the ampacity of the conductor for its trip setting. (Short circuit protection).

(a) 3 times (b) 5 times (c) 6 times (d) 8 times

49. Which of the following locations is **not** permitted for the use of surface metal raceways?

(a) dry location (b) hoistways (c) under raised floors (d) hazardous

50. In a Class I, Division II location a conduit passing through into a nonhazardous location, the sealing fitting shall be permitted ____.

(a) no seal required under the conditions (b) on either side of the boundary
(c) on both sides of the boundary (d) at the first fitting

51. Underfloor raceways not more than 100 millimeters wide shall have no less than ____ concrete or wood above them.

(a) 1/2 inch (b) 3/4 inch (c) 2 inches (d) 4 inches

52. A #250 kcmil conductor leaves through the wall opposite its removable and lay-in wire terminal intended for only one wire. The minimum bending space would be ____.

(a) 4" (b) 4 1/2" (c) 5" (d) 6 1/2"

53. A community antenna system grounding conductor shall not be smaller than ____ AWG, and shall have a current-carrying capacity approximately equal to that of the outer conductor of the coaxial cable.

(a) #18 (b) #16 (c) #14 (d) #12

54. Equipment installed and likely to become energized shall be grounded at which of the following distances?

(a) 8 feet horizontal and 5 feet vertical of grounded metal objects
(b) 5 feet vertical and 5 feet horizontal of grounded metal objects
(c) 8 feet vertical and 5 feet horizontal of grounded metal objects
(d) 8 feet vertical and 8 feet horizontal of grounded metal objects

55. Boxes having an approved system of organic coatings and are installed out of doors shall be marked ____.

(a) weatherproof (b) raintight (c) watertight (d) outdoor usage

56. A receptacle installed outdoors shall be considered protected from the weather by being located under ____ and will not be subjected to a beatung rain or water run off.

I. roofed open porches
II. canopies
III. marquees

(a) I only (b) II and III only (c) I and II only (d) I, II and III

57. A cutout box installed in a wet location shall be ____.

(a) raintight (b) weatherproof (c) waterproof (d) rainproof

58. Electrical nonmetallic tubing is permitted ____.

I. concealed in walls, floors and ceilings with a 15 minute fire rating
II. embedded in concrete provided with approved fitting
III. directly buried
IV. above a suspended ceiling with a 15 minute fire rating

(a) I only (b) I, II and IV (c) I, II and III (d) all of the above

59. Where load is connected to the tie at any point between transformer supply points and overcurrent protection is not provided in accordance with Article 240, the rated ampacity of the tie shall not be less than ___ percent of the rated secondary current of the largest transformer connected to the secondary tie system.

(a) 100 (b) 150 (c) 250 (d) 300

60. The branch circuit rating of a non-motor operated appliance that is continuously loaded shall have a minimum rating of ____ percent of the marked rating.

(a) 50 (b) 80 (c) 125 (d) 150

61. Each service disconnecting means shall ____ disconnect all ungrounded service conductors from the premises wiring system.

(a) automatically (b) independently (c) simultaneously (d) separately

62. A conductor at a terminal which leaves a cabinet, the minimum wire bending space at the terminal shall be ____ inches if the conductor size is #2 and it does not enter the enclosure through the wall opposite.

(a) 2 (b) 2 1/2 (c) 3 1/2 (d) 4

63. A 50 volt generator which is driven by a single motor is protected by the overcurrent protecting the motor only when the generator is delivering no more than ____ percent of its full load rated current.

(a) 80 (b) 100 (c) 125 (d) 150

64. MI cable shall be permitted ____.

I. as feeders and branch circuits
II. for wet and dry locations
III. for concealed or exposed

(a) I only (b) II only (c) III only (d) I, II and III

65. If SE or USE cable consists of two or more conductors, one shall be permitted to be ____.

(a) insulated (b) green (c) tagged (d) uninsulated

66. For small motors not covered by the motor tables, the locked-rotor current shall be assumed to be ____ times the full load current.

(a) three
(b) four
(c) five
(d) six

67. Fuses and circuit breakers shall be so located or ____ that persons will not be burned or otherwise injured by their operation.

(a) concealed (b) guarded (c) shielded (d) elevated

68. Surface marking of conductors and cables shall be durably marked on the surface at intervals not exceeding ____ inches.

(a) 6 (b) 12 (c) 18 (d) 24

69. The cord shall not exceed ____ feet in length for portable or mobile signs in dry locations.

(a) 6 (b) 8 (c) 10 (d) 15

70. The horsepower rating of the disconnecting means of hermetic refrigerant motor-compressors shall be determined from the ____.

I. nameplate rated-load current
II. branch-circuit selection current
III. loads

(a) I only (b) II only (c) III only (d) I or II only

2005

OPEN BOOK EXAM #9

70 QUESTIONS
TIME LIMIT - 2 HOURS

TIME SPENT ______ **MINUTES**

SCORE ______ **%**

MASTER OPEN BOOK EXAM #9 Two Hour Time Limit

1. AC circuits of less than 50 volts shall be grounded under which of the following?

I. Where installed as overhead conductors outside of buildings.
II. Where supplied by transformers if the transformer supply system is ungrounded.
III. Where supplied by transformers if the transformer supply system exceeds 150 volts to ground.

(a) I only (b) II only (c) III only (d) I, II or III

2. Which of the following conditions must be met for vehicle mounted generators used to supply electrical tools on a construction site?

I. The frame of the generator is bonded to the vehicle frame.
II. The generator supplies cord and plug connected equipment thru receptacles mounted on the vehicle or on the generator.

(a). I only (b) II only (c) both I and II (d) neither I nor II

3. Distribution systems for mobile home parks shall be ____.

(a) 120/240v three-phase (b) 120/208v three-phase
(c) 120/240v single-phase (d) 120/230v single-phase

4. Metal oxide surge arrester ratings are based on the magnitude and duration of overvoltage at the arrester location as affected by ____.

I. switching surges II. system grounding techniques III. phase-to-ground faults

(a) I only (b) II only (c) III only (d) I, II and III

5. ____ devices providing equivalent overcurrent protection in closed-loop power distribution systems shall be permitted as a substitute for fuses or circuit breakers.

(a) Approved (b) Listed (c) Accessible (d) Automatic

6. A unit or assembly of units or sections, and associated fittings, forming a rigid structural system used to securely fasten or support cables and raceways is a ____.

(a) flat cable assembly (b) wireway (c) multioutlet assembly (d) cable tray system

7. ____ is the distance measured along the enclosure wall from the axis of the centerline of the terminal to a line passing through the center of the opening in the enclosure.

(a) Offset (b) Radius (c) Center point (d) none of these

8. Which of the following is permitted to support a lighting fixture weighing over 50 pounds?

(a) The screwshell of a lampholder.
(b) A box or fitting that is listed for the weight to be supported.
(c) An outlet box.
(d) Fixture wires #14 and larger.

9. Which of the following statements about junction boxes is/are true?

I. All shall have a cover II. All over 6' in length shall have conductors cabled or racked

(a) I only (b) II only (c) both I and II (d) neither I nor II

10. Branch circuits in a recreational vehicle may be derived from autotransformers if ____.

I. the transformer is listed for vehicle use II. UL and CSA listed and approved

(a) I only (b) II only (c) both I and II (d) neither I nor II

11. All of the following about paralleling conductors are true except ____.

(a) must terminate in the same manner **(b) must be same material**
(c) must be same length **(d) must be enclosed in the same raceway**

12. MI cable, bends in the radius of the inner edge of any bend shall not be less than ____ times the external diameter of the metallic sheath for cable not more than 3/4" in external diameter.

(a) 5 (b) 6 (c) 7 (d) 8

13. All of the following may be used on services of 2300/4600v except ____.

(a) MI cable (b) MV cable (c) cable bus (d) busway

14. Nonpower-limited fire protective signaling circuits shall ____.

I. not be more than 600 volts
II. not exceed 7 amps overcurrent protection for #18 conductor
III. be permitted in same raceway whether AC or DC current

(a) I only (b) I and II only (c) II only (d) I, II and III

15. Wire mesh or other conductive elements where provided in the concrete floor of animal confinement areas to provide an equipotential plane shall be bonded to the building grounding electrode system. The bonding conductor shall be copper, insulated, covered or bare, not smaller than # ____.

(a) 8 (b) 6 (c) 4 (d) 2

16. The provisions of section 300.20 shall not apply to the installation of a ____ conductor in a ferromagnetic envelope (metal enclosure).

(a) grounded (b) single (c) insulated (d) buried

17. A transformer vault ventilation opening shall be located ____ from doors, windows, fire escapes, and combustible material.

(a) 3 feet (b) 4 feet (c) 4 1/2 feet (d) as far away as possible

18. Concealed knob-and-tube wiring shall not be used in the hollow spaces of walls, ceilings and attics where such spaces ____.

(a) exceed 30 degrees C
(b) are insulated by loose or rolled insulation material
(c) are not fire rated for 3 hours
(d) are not ventilated

19. The supply cord conductors and internal wiring of portable high-pressure spray washing machines shall have ____ protection for personnel.

(a) thermal overloads **(b) current limiting fuses**
(c) factory installed GFCI **(d) inverse-time breakers**

20. The bending radius for nonshielded cables operating at 4160v shall be less than ____ times the diameter.

(a) 6 (b) 8 (c) 10 (d) 12

21. This article covers the installation and wiring of separately derived systems operating at 120 volts line-to-line and ____ volts to ground for senstitive electronic equipment.

(a) 12 (b) 24 (c) 50 (d) 60

22. Fixed electric space heating equipment requiring supply conductors with over ____ insulation shall be clearly and permanently marked.

(a) 60° F (b) 75° C (c) 60° C (d) 90° C

23. Totally enclosed motors of Types (2) or (3) shall have no external surface with an operating temperature in degrees Celsius in excess of ____ of the ignition temperature of the gas or vapor involved.

(a) 100% contained (b) 80% (c) 75% (d) none of these

24. Listed Christmas tree and decorative lighting outfits shall be permitted to be smaller than ____.

(a) #18 AWG (b) #20 AWG (c) #22 AWG (d) #26 AWG

25. All boxes and enclosures for emergency circuits shall be marked so they will be ____ as a component of an emergency circuit.

(a) readily identified (b) recognized (c) easily sighted (d) classified

26. The rating of an adjustable trip circuit breaker having ____ means for adjusting the current setting (long-time pickup setting), not meeting the requirements of (c), shall be the maximum setting possible.

(a) external (b) an isolated (c) readily accessible external (d) accessible

27. Each section, panel, or strip carrying a number of infrared lampholders shall be considered a(an) ____.

(a) light fixture (b) appliance (c) receptacle (d) outlet

28. This type equipment shall carry a prominent, permanently installed warning regarding the necessity for this grounding feature.

(a) Class I, Division I motor equipment
(b) electrostatic equipment
(c) all service entrance equipment in excess of 1200 amps
(d) Class II, Division I service equipment

29. Only wiring methods consisting of ____ shall be installed in ducts or plenums used for environmental air.

I. EMT II. type NMC III. type MI IV. flexible metallic tubing

(a) I and II only (b) I, II and III only (c) I, III and IV only (d) I, II, III and IV

30. It is desirable to limit the size of the isolation transformer to ____ kva or less and to use conductor insulation with low leakage to meet impedance requirements.

(a) 10 (b) 15 (c) 20 (d) 30

31. For emergency systems, the authority having jurisdiction shall conduct or witness a test on the complete system upon installation and periodically afterward. A ____ shall be kept of such tests.

(a) report (b) log (c) written record (d) chart

32. The service conductors shall be connected to the service disconnecting means by ____ or other approved means.

I. clamps II. pressure connectors

(a) I only (b) II only (c) both I and II (d) neither I nor II

33. Individual showcases, other than fixed, shall be permitted to be connected by flexible cord to permanently installed receptacles. The installation shall comply with which of the following?

I. the wiring will not be exposed to mechanical damage
II. attachment plugs shall be of a listed grounding type
III. flexible cord shall be hard-service type

(a) I only (b) II only (c) III only (d) I, II and III

34. The grounded conductor of a mineral-insulated, metal-sheathed cable shall be identified at the time of installation by ____ marking at its termination.

(a) distinctive (b) neutral (c) solid (d) identified

35. The prime mover of an emergency generator set ____.

I. must be provided with an automatic means for starting
II. must be provided with an automatic means of transferring from one fuel supply to another, where dual supplies are used.
III. must have an on-site fuel supply sufficient to operate the prime mover at full demand for 2 hours

(a) I only (b) II only (c) III only (d) I, II and III

36. Where a service mast is used for the support of service drop conductors, it shall be of adequate strength or be supported by ____.

(a) studs (b) braces or guys (c) rigid conduit (d) R.C. beams

37. For open conductors of not over 600 volts, clearance from ground shall be ____ feet, above finished grade where the supply conductors have a nominal voltage limited to 150 volts to ground and accessible to pedestrians only.

(a) 10 (b) 12 (c) 8 (d) 15

38. The following pool equipment shall be grounded ____.

I. ground-fault circuit-interrupters
II. transformer enclosures
III. electric equipment located within 5 feet of the inside wall of the pool

(a) III only (b) II and III only (c) II only (d) I, II and III

39. The minimum wire bending space within the enclosure for motor controllers for a #1/0 conductor (2 wires per terminal) would be ____ inches.

(a) 6 (b) 5 (c) 4 (d) 2

40. It is the intent of the Code that ____ wiring or the construction of equipment need not be inspected at the time of installation of the equipment, if the equipment has been listed by a qualified electrical testing laboratory.

(a) factory-installed internal **(b) factory-installed**
(c) underground **(d) raceway**

41. Any motor application shall be considered as ____ unless the nature of the apparatus it drives is such that the motor will not operate continuously with load under any condition of use.

(a) short-time duty **(b) varying duty**
(c) continuous duty **(d) periodic duty**

42. An overcurrent trip unit of a circuit shall be connected in series with each ____.

(a) ungrounded conductor **(b) grounded conductor**
(c) overcurrent device **(d) transformer**

43. Range hoods shall be permitted to be cord-and-plug connected with a flexible cord identified as suitable for use on range hoods in the installation instructions of the appliance manufacturer, if which of the following conditions are met?

I. The receptacle is readily accessible.
II. The receptacle is supplied by an individual branch circuit.
III. Receptacles are located to avoid physical damage to the flexible cord.

(a) I, II, and III (b) I and II only (c) II and III only (d) I and III only

44. Traveling cables shall be supported by looping the cables around supports for unsupported lengths less than ____ feet.

(a) 300 (b) 200 (c) 150 (d) 100

45. Optional standby systems are ____.

(a) covered under Article 700 emergency systems
(b) not covered in the Code
(c) covered under Article 701
(d) covered in their own section

46. In all cases the work space in front of electrical equipment shall permit at least a ____ degree opening of equipment doors or hinged panels.

(a) 60 (b) 90 (c) 120 (d) 180

47. The N.E.C. covers ____.

I. electronic organs II. speech-input systems III. audio signal generation

(a) I only (b) II only (c) III only (d) I, II and III

48. The protective devices(s) shall be capable of detecting and interrupting all values of current which can occur at their location in excess of their trip setting or ____.

I. boiling point II. melting point III. capacity

(a) I only (b) II only (c) III only (d) I, II and III

49. A system (of two-wire DC) operating at 50v or less between conductors shall ____.

(a) not be grounded **(b) be grounded**
(c) not be permitted **(d) not be required to be grounded**

50. Mobile x-ray equipment is mounted on a ____ base with wheels and/or casters for moving while completely assembled.

(a) portable (b) transportable (c) permanent (d) temporary

51. It is desirable to locate the ammeter so that it is conspicuously visible to persons in the ____ location.

(a) Class I (b) Class II (c) Class III (d) anesthetizing

52. An askarel-insulated transformer installed in a poorly ventilated place shall be furnished with ____.

I. a means for absorbing any gases generated by arcing inside the case
II. the pressure-relief vent shall be connected to a flue that will carry such gases outside the building
III. the pressure-relief vent shall be connected to a chimney that will carry such gases outside the building

(a) I or II only (b) I or III only (c) I only (d) I, II or III

53. Storage batteries for diesel engine drives for fire pumps shall be ____ above the floor.

(a) 4' (b) 5' (c) 6' (d) rack supported

54. The terminals of an electric-discharge lamp shall be considered energized where any lamp terminal is connected to a circuit of over ____ volts.

(a) 100 (b) 200 (c) 300 (d) 500

55. Legally required standby systems are typically installed to serve loads such as ____.

I. sewerage disposal II. data processing III. refrigeration systems

(a) I and II only (b) II and III only (c) I and III only (d) I, II and III

56. Cases or frames of current transformers, the primaries of which are not over 150 volts to ground and which are used exclusively to supply current to meters ____.

(a) need to be grounded (b) need to be isolated
(c) need to be insulated (d) need not be grounded

57. ____ shall be permitted to be installed in concrete, in direct contact with the earth, or in areas subject to severe corrosive influences where protected by corrosion protection and judged suitable for the condition.

(a) PVC (b) Ceramic (c) Orangeburg (d) Rigid metal conduit

58. Connections from headers to cabinets and other enclosures in cellular concrete floor raceways, shall be made by means of ____ raceways and approved fittings.

(a) rigid nonmetallic (b) metal (c) non-metallic (d) all of these

59. All heating elements that are ____ and are a part of an electric heater shall be legibly marked with the ratings in volts and watts, or in volts and amperes.

I. rated over one amp II. replaceable in the field III. a part of an appliance

(a) I only (b) II only (c) III only (d) I, II and III

60. Type ITC cable is permitted to be used in industrial establishments with proper supervision in which of the following?

I. aerial cable on a messenger II. cable trays III. with nonpower-limited circuits

(a) I only (b) II only (c) I and II only (d) I, II and III

61. Distances from signs, radio, and TV antennas, tanks or other nonbuilding or nonbridge structures, clearances, vertical, diagonal and horizontal, shall not be less than ____ feet.

(a) 2 (b) 3 (c) 6 (d) 8

62. A 1 1/2" rigid metal conduit, 30" long containing over 2 conductors may be filled to ____ square inch.

(a) 1.224 (b) 1.122 (c) 0.829 (d) 0.6324

63. You are to install a 1/2" flexible metallic tubing in an area so that after installation its use will be infrequent flexing, the radius of bends shall not be less than ____ inches.

(a) 3 1/2 (b) 4 (c) 12 1/2 (d) 17 1/2

64. Circuit breakers shall be marked with a voltage rating no less than the nominal system voltage that is indicative of their capability to interrupt ___.

I. fault currents between phases
II. fault currents from phase to ground
III. fault currents between the neutral and ground

(a) I only (b) I & II only (c) II & III only (d) I, II, or III

65. A Class 2 circuit considers safety _____.

I. provides acceptable protection from electric shock
II. from a fire initiation standpoint

(a) I only (b) II only (c) both I and II (d) neither I nor II

66. The frames of stationary motors are grounded under which of the following conditions?

I. where supplied by metal-enclosed wiring
II. in a wet location and not isolated or guarded
III. effectively isolated from the ground
IV. if the motor operates with any terminal at over 150 volts to ground

(a) I and II only (b) I, II and IV only (c) III only (d) I, II, III and IV

67. A/an ____ is a device that changes DC input to AC output.

I. inverter II. power conditioning unit III. power conversion system

(a) I only (b) II only (c) III only (d) I, II or III

68. In a 1500 sq.ft. house, the minimum feeder neutral load for a 5 kva clothes washer/dryer would be ____ kva.

(a) 5.0 (b) 4.3 (c) 3.5 (d) 3.0

69. Formal interpretations of the Code may be found in the ____.

(a) National Electrical Code Handbook
(b) OSHA Standards
(c) NFPA Regulations Governing Committee Projects
(d) Life and Safety Handbook

70. The allowable area to be filled of a 2" EMT conduit is ____ sq.in., if it contains 3 or more conductors.

(a) 2.067 (b) 3.36 (c) 0.8268 (d) 1.342

2005

OPEN BOOK EXAM #10

70 QUESTIONS
TIME LIMIT - 2 HOURS

TIME SPENT ☐ **MINUTES**

SCORE ☐ **%**

MASTER OPEN BOOK EXAM #10 Two Hour Time Limit

1. Where NM cable or underground feeder and branch circuit cable is used with boxes no larger than a nominal size 2 1/4" x 4" mounted in walls and where the cable is fastened within ____ inches of the box measured along the sheath and where the sheath extends into the box no less than 1/4", securing the cable to the box shall not be required.

(a) 8 (b) 10 (c) 12 (d) 24

2. A bare copper conductor can be used in an underground service ____.

(a) where judged suitable for soil conditions
(b) where used in a raceway
(c) without regard to soil conditions where part of cable assembly identified for underground use
(d) all of the above

3. The grounding conductor for communication circuits shall be connected to the ____.

I. metallic power raceway II. service equipment enclosure III. building electrode system

(a) I only (b) II only (c) II or III only (d) I, II or III

4. "Direct grade level access" is defined as being located not more than ____ above grade level.

(a) 6 1/2' (b) 6' 3" (c) 6' 4" (d) 8'

5. A main bonding jumper shall be a ____.

I. wire II. screw III. bus

(a) I only (b) I or II only (c) I or III only (d) I, II or III

6. The GFCI protection device shall be located not less than ____ above the established electrical datum plane.

(a) 3" (b) 6" (c) 10" (d) 12"

(a) equipment grounding (b) grounded (c) neutral (d) bonding

7. Each receptacle of DC plugging boxes shall be rated at not less than ____ amps.

(a) 15 (b) 20 (c) 25 (d) 30

8. A protective layer which is installed between the floor and type FCC flat conductor cable to protect the cable from physical damage and may or may not be incorporated as an integral part of the cable is the ____.

(a) transition assembly (b) outer sheath (c) bottom shield (d) header

9. An area must be classed as a Class II hazardous location if it contains ____.

(a) combustible dust (b) ignitable vapors (c) flammable gases (d) ignitable fibers

10. The minimum headroom requirement of 6' 6" for service equipment in an existing dwelling unit is not required if the service equipment or panelboard does not exceed ____ amps.

(a) 100 (b) 200 (c) 300 (d) 400

11. Which of the following statements about lightning protection is (are) true?

I. Air terminals should be spaced at least 6 feet away from any other electrode.
II. If lightning protection is required for an irrigation machine, a driven ground rod should be connected to the machine at a stationary point.

(a) I only (b) II only (c) both I and II (d) neither I nor II

12. If the appliance is provided with a single-pole switching device, the attachment plug shall be ____.

I. of the grounding type II. polarized

(a) I only (b) II only (c) I or II (d) neither I nor II

13. Listed equipment protected by a system of double insulation, or its equivalent, shall not be required to be grounded. Where such a system is employed, the equipment shall be ____.

(a) labeled (b) approved (c) distinctively marked (d) identified

14. The ampacity of a #12 THW conductor used with a 15-minute motor on a monorail hoist installed in a raceway containing a total of four current-carrying conductors would be ____ amps.

(a) 20 (b) 30 (c) 33 (d) 36.96

15. No premises wiring, with a grounded conductor, shall be electrically connected to a supply system unless the supply system contains ____.

(a) a grounded conductor
(b) a wiring design
(c) protection
(d) none of these

16. A manufactured wiring system shall have receptacles that are ____.

I. uniquely polarized II. of the locking type III. GFCI

(a) I only (b) II only (c) I and II only (d) I, II and III

17. For small motors the locked-rotor current shall be assumed to be ____ the full-load current.

(a) 4 (b) 5 (c) 6 (d) 8

18. A cord connector that is supported by a permanently installed cord pendant shall be considered a(an) ____.

(a) receptacle outlet (b) permanent cord (c) lighting outlet (d) outlet device

19. Proscenium side lights shall have the lampholder terminals kept away at least ____ from the metal of the trough.

(a) 1/2" (b) 1" (c) 3" (d) 6"

20. Stage cables used in motion picture studios for stage lighting shall be protected by means of overcurrent devices set at not more than ____ of the values given in the appropriate Code table.

(a) 500% (b) 400% (c) 200% (d) 100%

21. When installing a surge arrester at the service of less than 1000 volts, the grounding conductor shall be connected to ___.

I. the grounded service conductor
II. the grounding electrode conductor
III. the grounding electrode for the service
IV. the equipment grounding terminal in the service equipment

(a) I and II only (b) I and III only (c) III and IV only (d) I, II, III, or IV

22. ____ of conductors in rigid nonmetallic conduit shall be made only in junction, outlet boxes or conduit bodies.

(a) Splices (b) Splices and taps (c) Connections (d) none of these

23. Connection of conductors to terminal parts shall ensure a thoroughly good connection without damaging the conductors and shall be made by means of ____.

I. splices to flexible leads II. pressure connectors III. solder lugs

(a) I only (b) II only (c) I and II only (d) I, II or III

24. ____ of insulating material shall be permitted to be used without boxes in exposed cable wiring.

I. Switch devices II. Outlet devices III. Tap devices

(a) I only (b) II only (c) III only (d) I, II and III

25. A nursing home is a building or part thereof used for the lodging, boarding and nursing care, on a 24-hour basis, of ____ or more persons.

(a) 4 (b) 12 (c) 50 (d) 100

26. Which of the following must be provided with GFCI?

(a) dishwashers (b) fountains (c) outdoor lights (d) refrigerators

27. Each resistance welder shall have overcurrent primary protection set at not more than ____ percent.

(a) 200 (b) 300 (c) 250 (d) 125

28. 3-way and 4-way switches shall be so wired that all switching is done in the ____ conductor.

(a) ungrounded (b) grounded (c) neutral (d) grounding

29. Service conductors on a 4600v system shall not be smaller than ____ unless in cable.

(a) #6 (b) #4 (c) #2 (d) #1

30. In cellular metal floor raceways all of the following are true **except** ____.

(a) splices and taps can be made in junction boxes
(b) disconnected outlets are removed
(c) entry boxes are installed flush to the floor
(d) the combined cross sectional fill cannot exceed 45%

31. Non-shielded high-voltage cables shall be installed in ____ conduit encased in not less than 3" of concrete.

I. rigid PVC   II. IMC   III. rigid metal

(a) I only   (b) II only   (c) III only   (d) I, II or III

32. On solar photovoltaic system; Ampacity of conductors and overcurrent devices shall not be less than ____ percent of the computed current.

(a) 150   (b) 100   (c) 125   (d) 200

33. A run of flexible metal conduit may be used as an equipment grounding conductor if the conductors are protected at ____.

(a) 20a or more   (b) 20a or less   (c) 30a or more   (d) 30a or less

34. For devices with screw shells, the terminal for the ____ conductor shall be the one connected to the screw shell.

(a) green   (b) grounded   (c) ungrounded   (d) grounding

35. The service to a mobile home may consist of ____.

I. 3 - 50a separate services   II. one permanently installed feeder

(a) I only   (b) II only   (c) both I and II   (d) neither I nor II

36. A motor is connected to the electrical system by #10 THW conductors that are protected by a double pole 40 amp breaker. These conductors are enclosed in a 1/2" liquidtight flexible metal conduit that is 4 feet long and is terminated in connectors that are approved for grounding. Which of the following is the correct statement concerning this installation?

(a) The 1/2" liquidtight conduit will not permit the installation of two #10 THW's due to the 40% fill requirement.
(b) The conduit must contain a grounding conductor since it is more than 3' in length.
(c) The conduit must contain a grounding conductor since the conductors it contains are fused at more than 20 amps.
(d) The conduit must contain a grounding conductor since it contains conductors that are connected to a motor load.

37. You are installing a 75 foot run of 2" rigid metal conduit, using threaded couplings. The Code requires you to support this conduit within 3 feet of termination, and then a maximum of ____ feet apart.

(a) 3 (b) 8 (c) 12 (d) 16

38. The #4 solid copper conductor connecting the grounding electrode to the service has been accidentally cut. Which of the following would be acceptable connectors for splicing this conductor?

I. a split bolt connector
II. a compression sleeve type connector
III. a wire nut

(a) I only (b) I or II only (c) I, II or III (d) none of these are acceptable

39. In panelboards, where the voltage on busbars is 150 volts and the bars are opposite polarity, held free in air, the minimum spacing between the parts is ____.

(a) 3/4" (b) 1" (c) 1 1/2" (d) 2"

40. ____ is a combination consisting of a compressor and motor, both of which are enclosed in the same housing, with no external shaft or shaft seals, the motor operating in the refrigerant.

I. Motor-compressor
II. Hermetic refrigerant motor-compressor
III. Air-conditioning equipment

(a) I only (b) II only (c) III only (d) I, II and III

41. A 3" conduit, not more than ____ inches in length, connecting a gutter and a switch case, may have a conductor fill of 60% of the internal cross-sectional area.

(a) 6 (b) 12 (c) 18 (d) 24

42. The service entrance conductors are #350 kcmil copper. The minimum size grounding electrode conductor allowed by the Code would be a ____ copper, if it were the sole connection to a made electrode.

(a) #4 (b) #2 (c) #6 (d) #8

43. Supplementary overcurrent devices shall not be required to be ____.

(a) accessible (b) readily accessible (c) continuous duty (d) adjustable

44. Where single conductors are installed in a triangular or square configuration in uncovered cable trays, with a maintained free air space of not less than ____ times one conductor diameter of the largest conductor contained within the configuration and adjacent conductor configurations, the ampacity of #1/0 AWG and larger cables shall not exceed the allowable ampacities of two or three single insulated conductors rated 0 through 2000 volts supported on a messenger.

(a) 3 (b) 2 (c) 2.15 (d) 2.75

45. All dwelling unit 125v, single-phase 15 and (or) 20 amp receptacles installed in a ____ shall have GFCI protection for personnel.

I. bathroom
II. crawl space at or below grade level
III. bedroom

(a) I and II only (b) II and III only (c) II only (d) I, II and III

46. All electric spa or hot tub water heaters shall be listed and have the heating elements subdivided into loads not exceeding ____ amperes and protected at not more than ____ amperes.

(a) 45 ... 50 (b) 48 ... 60 (c) 40 ... 45 (d) 55 ... 60

47. Capacitors shall be permitted to be protected ____.

I. in groups II. individually

(a) I only (b) II only (c) I or II (d) neither I nor II

48. Listed ceiling fans that do not exceed ____ pounds in weight, with or without accessories, shall be permitted to be supported by outlet boxes.

(a) 35 (b) 40 (c) 45 (d) 50

49. When service entrance phase conductors are larger than 1100 kcmil copper, the bonding jumper shall have an area not less than what percent of the area of the largest phase conductor?

(a) 6% (b) 10% (c) 12 1/2% (d) 15%

50. Surface-mounted fluorescent lighting fixtures that contain a ballast and are to be installed on combustible, low-density cellulose fiberboard should be spaced not less than ____ inches from the surface of the fiberboard.

(a) 1/4" (b) 1/2" (c) 1" (d) 1 1/2"

51. A 15 or 20 amp rated receptacle outlet shall be installed at an accessible location for the servicing of heating, A/C and refrigeration equipment. The receptacle outlet shall be located on the same level and within ____ feet of the equipment.

(a) 25 (b) 50 (c) 75 (d) 100

52. The outer metal shell of a lampholder shall be lined with insulating material that shall prevent the shell and cap from becoming a part of the circuit. The lining shall not extend beyond the metal shell more than ___, but shall prevent any current carrying part of the lamp base from being exposed when a lamp is in the lampholding device.

(a) 1/16" (b) 1/8" (c) 1/4" (d) 1/2"

53. Where shore power accommodations provide two receptacles specifically for an individual boat slip and these receptacles have different voltages, only the receptacle with the _____ shall be required to be calculated.

(a) highest voltage (b) lowest voltage (c) highest amperage (d) larger kilowatt demand

54. Articles ____ through ____ cover occupancies or parts of occupancies that are or may be hazardous because of atmospheric concentrations of flammable liquids, gases, or vapors, or because of deposits or accumulations of materials that may be readily ignitible.

(a) 517 through 520 (b) 511 through 517 (c) 514 through 517 (d) 514 through 600

55. Isolating switches over 600v shall be provided with a means for readily connecting the load side conductors to ground when disconnected from the ____.

(a) current (b) equipment (c) service cable (d) source of supply

56. Time switches, flashers, and similar devices where mounted so they are accessible only to qualified persons shall be permitted without barriers, provided they are located within an enclosure such that any energized part within ____ of the manual adjustment or switch are covered by suitable barriers.

(a) 4" (b) 6" (c) 12" (d) 18"

57. Omission of overcurrent protection shall be permitted at points where busways are reduced in size, provided that the smaller busway doesn't extend more than ____ feet.

(a) 10 (b) 20 (c) 50 (d) 70

58. A cellular concrete floor raceway's grounding conductor shall connect the insert receptacles to a positive ground connection provided on the ____.

(a) junction box (b) cell (c) fitting (d) header

59. A circuit breaker shall be of such design that any alteration of its ____ will require dismantling of the device or breaking of a seal for other than intended adjustments.

I. trip point II. time required for its operation

(a) I only (b) II only (c) both I and II (d) neither I nor II

60. All of the following motors are permitted in a Class III, Division I area except ____.

(a) totally enclosed pipe ventilated **(b) non-ventilated**
(c) totally enclosed fan cooled **(d) water cooled**

61. The maximum overcurrent device on a branch circuit supplying a ASME rated boiler is ____ amps.

(a) 40 (b) 60 (c) 100 (d) 150

62. A device supplying running overload protection may be shunted during starting a motor when it is started ____.

I. manually II. automatically

(a) I only (b) II only (c) both I and II (d) neither I nor II

63. The size of branch-circuit conductors and overcurrent protective devices for electrode-type boilers shall be calculated on the basis of ____ percent of the total load (motors not included).

(a) 25 (b) 75 (c) 100 (d) 125

64. Which of the following statements is/are true?

I. On a grounded service, the grounded service neutral shall not be smaller than the grounding electrode conductor.
II. If a 1000v or less system is grounded, the grounded conductor must be run to each service.

(a) I only (b) II only (c) both I and II (d) neither I nor II

65. Maximum voltage between conductors serving a submersible pump in a fountain is ____ volts.

(a) 150 (b) 250 (c) 300 (d) 600

66. Which of the following about an aircraft hangar is true?

I. Any area below the floor level shall be considered a Class I, Division I location up to the floor level.
II. The area within 5' horizontally of aircraft power plants or fuel tanks shall be considered a Class I, Division II location extending from the floor to a level 5' above the upper surface of wings and engine enclosures.

(a) I only (b) II only (c) both I and II (d) neither I nor II

67. All of the following are true about transformer vaults **except** ____.

(a) ventilated openings are calculated at 3 sq.in. per kva
(b) each doorway shall be fully louvered for ventilation
(c) door sills and curbs shall not be less than 4" in height
(d) materials shall not be stored inside the vaults

68. The ____ and the bridge frame shall be considered as electrically grounded through the bridge and trolley wheels and its respective tracks unless local conditions, such as paint or other insulating material, prevent reliable metal-to-metal contact.

(a) trolley frame (b) track (c) trolley wheels (d) none of these

69. All but which of the following shall be continuous between cabinets, boxes, fittings or other enclosures or outlets?

(a) short sections of raceways used to provide support or protection of cable assemblies
(b) metallic or non-metallic raceways
(c) cable armors
(d) cable sheaths

70. A buried cable, 4160v, requires a minimum depth of ____ inches.

(a) 24 (b) 30 (c) 36 (d) 42

2005

OPEN BOOK EXAM #11

70 QUESTIONS
TIME LIMIT - 2 HOURS

TIME SPENT ☐ **MINUTES**

SCORE ☐ **%**

MASTER OPEN BOOK EXAM #11 **Two Hour Time Limit**

1. The minimum size conductor permitted in parallel for elevator lighting is ____, provided the ampacity is equivalent to a #14 wire.

(a) #14 (b) #20 (c) #16 (d) #1/0

2. Conductors of light and power systems of all voltages may occupy the same enclosure or raceway ____.

(a) if less than 600 volts and if insulated for maximum voltage of any conductor within the enclosure or raceway
(b) if power system is over 600 volts and light system under 600 volts
(c) if power system is over 600 volts and individual circuits are AC
(d) in most instances without qualification

3. Transformers rated over ____ KV shall be installed in a vault.

(a) 10 (b) 12 1/2 (c) 25 (d) 35

4. The neutral of feeders supplying solid-state, 3-phase, 4-wire dimming systems shall be considered a ____ conductor.

(a) current-carrying (b) noncurrent-carrying (c) balanced (d) isolated

5. Tubing having cut threads and used as arms or stems on light fixtures may not be less than ____ inches wall thickness.

(a) .020 (b) .025 (c) .040 (d) .015

6. Expansion fittings for nonmetallic wireway shall be provided to compensate for thermal expansion and contraction where the length change is expected to be ____ or greater in a straight run.

(a) .025 in. (b) .050 in. (c) .75 in (d) 1/4"

7. Types TPT, TS and TST shall be permitted in lengths not exceeding ____ feet when attached directly, or by means of a special type of plug, to a portable appliance rated at 50 watts or less.

(a) 8 (b) 10 (c) 15 (d) can't be used at all

8. When determining the load on the "volt-amps per square foot" basis, the floor area shall be computed from the ____ dimensions of the building.

(a) inside (b) outside (c) either (a) or (b) (d) neither (a) nor (b)

9. Parts of electric equipment which in ordinary operation produce ____ shall be enclosed or separated and isolated from all combustible material.

I. molten metal II. flames III. sparks

(a) II only (b) III only (c) I and II only (d) I, II and III

10. Locations where combustible dust is normally in heavy concentrations are designated as ____.

(a) Class I, Division II **(b) Class II, Division I**
(c) Class II, Division II **(d) Class III, Division I**

11. Which of the following is not a standard classification for a branch circuit supplying several loads?

(a) 20 amp (b) 25 amp (c) 30 amp (d) 50 amp

12. The minimum size conductor for lighting elevator circuits traveling cables is ____.

(a) #12 (b) #18 (c) #16 (d) #14

13. Steel cable trays shall not be used as equipment grounding conductors for circuits protected above ____ amperes.

(a) 200 (b) 60 (c) 600 (d) 1200

14. Conductive materials enclosing electrical conductors are grounded to ____.

I. prevent lightning surges
II. prevent voltage surges
III. facilitate the operation of the overcurrent device under ground-fault conditions

(a) I only (b) II only (c) III only (d) all of these

15. No grounded conductor shall be attached to any terminal or lead so as to reverse designated ____.

(a) phase (b) angle (c) polarity (d) line

16. Panelboards, switches, gutters, wireways or transformers are permitted to be mounted above or below one another if ___.

(a) rated 300v or less
(b) flush along the back edge
(c) they extend not more than 6 inches beyond the front of the equipment
(d) flush along the front edge

17. When the number of receptacles for an office building is unknown, an additional load of ____ volt-amp(s) per square foot is required.

(a) 1 (b) 2 (c) 1/2 (d) 3

18. If a protective device rating is marked on an appliance, the branch circuit overcurrent device rating shall not exceed ____ the protective device rating marked on the appliance.

(a) at all (b) more than 50% (c) 80% (d) 125%

19. Unless identified for use in the operating environment, no conductors or equipment shall be located in ____ having a deteriorating effect on the conductors or equipment.

I. damp or wet locations
II. where exposed to gases, fumes, vapors, liquids, or other agents

(a) I only (b) II only (c) I and II (d) none of these

20. A stage switchboard that is not completely enclosed dead-front and dead-rear or recessed into a wall shall be provided with a ____ extending the full length of the board to protect all equipment on the board from falling objects.

(a) cover (b) guard (c) mesh net (d) metal hood

21. The conduit or raceways, including their end fittings, shall not rise more than ____ inches above the bottom of the enclosure.

(a) 3 (b) 4 (c) 5 (d) 6

22. Service entrance cables, where subject to physical damage, shall be protected in ____.

I. EMT II. IMC III. rigid metal conduit

(a) III only (b) II and III (c) I, II and III (d) I and III

23. All buildings or portions of buildings or structures designed or intended as a place of assembly shall have ____ or more persons.

(a) 50 (b) 100 (c) 250 (d) 500

24. ____ and larger grounding electrode conductors shall be protected where exposed to severe physical damage.

(a) #8 (b) #4 (c) #2 (d) #6

25. Where storage batteries are used for emergency systems they shall be ____.

(a) provided with automatic battery charging means
(b) alkali or acid type
(c) neither (a) nor (b)
(d) both (a) and (b)

26. The following letter suffixes shall indicate the following:

____ for two insulated conductors laid parallel within an outer nonmetallic covering.

(a) D (b) M (c) R (d) N

27. When derating the ampacity of multiconductor cables to be installed in cable tray, the ampacity deration shall be based on ___.

I. the total number of current carrying conductors in the cable tray
II. the total number of current carrying conductors in the cable

(a) I only (b) II only (c) either I or II (d) both I and II

28. When connections are made in the white wire in a multi-wire circuit at receptacles, they are required to be made ____.

(a) connected to the silver terminals on the duplex
(b) to the brass colored terminal
(c) with a pigtail to the silver terminal
(d) none of these

29. In a grounded system, the conductor that connects the circuit grounded conductor at the service and/or the equipment grounding conductor to the grounding electrode is called the ____.

(a) main grounding conductor
(b) common main grounding conductor
(c) equipment grounding conductor
(d) grounding electrode conductor

30. Underground cable installed under a building shall be in a raceway that is ____.

(a) encased **(b) extended beyond the outside walls of a building**
(c) buried at least 6 inches **(d) not buried more than 12 inches**

31. The overall covering for type NMC cable shall be ____.

I. flame retardant
II. moisture resistant
III. fungus resistant
IV. corrosion resistant
V. all of these

(a) II and III (b) I and III (c) II, III and IV (d) V

32. Circuit breakers shall be so located or shielded so that persons ____.

(a) will not be burned or otherwise injured by their operation
(b) other than the authority cannot locate them
(c) cannot operate them without a key
(d) other than the authority cannot remove them

33. Which of the following are **not** classified patient care areas?

I. day rooms II. lounges III. business offices

(a) II only (b) II and III only (c) III only (d) I, II and III

34. The neutral feeder conductor must be capable of carrying the maximum ____ load.

(a) connected (b) unbalanced (c) demand (d) grounded

35. For straight pulls, the length of the box shall be not less than ____ the outside diameter, over sheath, of the largest conductor or cable entering the box on systems over 600 volts.

(a) 8 times (b) 6 times (c) 36 times (d) 48 times

36. Expansion joints and telescoping sections of raceway shall be made electrically continuous by equipment ____ or other means approved for the purpose.

(a) grounding conductors **(b) grounded conductor**
(c) bonding jumpers **(d) none of these**

37. Sealing compound is employed with mineral-insulated cable in a Class I location for the purpose of ____.

(a) preventing passage of gas or vapor
(b) excluding moisture
(c) limiting a possible explosion
(d) preventing escape of powder

38. A lighting and appliance branch circuit panelboard contains six - 3 pole circuit breakers and eight - 2 pole circuit breakers. The maximum allowable number of single pole circuit breakers permitted to be added is ____.

(a) 8 (b) 16 (c) 28 (d) 12

39. The Code provides that unshielded lead-in conductors of amateur transmitting stations shall clear the building surface which is wired over by a distance not less than ____ inches.

(a) 1 (b) 2 (c) 3 (d) 4

40. A 24 volt landscape lighting system installed with UF cable is permitted with a minimum cover of ____ inches.

(a) 6 (b) 12 (c) 18 (d) 24

41. Each circuit leading to or through a dispensing pump shall be provided with a switch or other acceptable means to disconnect ____ from the source of supply all conductors of the circuit, including the grounded conductor, if any.

(a) automatically (b) simultaneously (c) manually (d) individually

42. Floor boxes shall be considered to meet the requirement of the spacing receptacles on walls if they are ____.

(a) within 24" of the wall **(b) within 18" of the wall**
(c) close to the wall **(d) none of these**

43. For general wiring in Class I, Division I locations it is permissible to use ____.

(a) rigid metal conduit (b) EMT (c) flexible metal conduit (d) all of these

44. In closed construction in a manufactured building, cables shall be permitted to be secured only at cabinets, boxes, or fittings where ____ or smaller conductors are used and protected as required.

(a) #2 AWG (b) #10 AWG (c) #2/0 AWG (d) #250 kcmil

45. In a commercial garage the pit shall be classified ____ unless provisions are made for six air changes per hour.

(a) Class I, Division II
(b) Class II, Division II
(c) Class II, Division I
(d) Class I, Division I

46. Ground clamps shall be approved for general use without protection or shall be protected ____.

I. by enclosing in wood
II. by enclosing in metal
III. by equivalent protective covering

(a) II only (b) II and III only (c) I and III only (d) I, II and III

47. Each doorway leading into a vault from the building interior shall be provided with a tight fitting door having a minimum fire rating of ____ hours.

(a) 2 (b) 4 (c) 5 (d) 3

48. Auxiliary equipment for electric-discharge lamps shall be enclosed in noncombustible cases and ____.

(a) not over 3" away **(b) not over 1500w**
(c) treated as sources of heat **(d) none of these**

49. Tap conductors for household cooking equipment supplied from a 50 amp branch circuit shall have an ampacity of not less than ____.

(a) 50 (b) 70 (c) 20 (d) 80

50. The ampacity for the supply conductors for a resistance welder with a duty cycle of 15% and a primary current of 21 amps is ____ amps.

(a) 9.45 (b) 8.19 (c) 6.72 (d) 5.67

51. Insulated wires shall be marked or tagged with which of the following?

(a) maximum rated voltage
(b) proper type letters
(c) manufacturer indentification
(d) all of these

52. A bonding jumper shall be used to connect the equipment grounding conductors of the derived system to the grounded conductor. This connection shall be made ___.

I. at any point on the separately derived system from the source to the first system disconnect
II. at any point on the separately derived system from the source to the first overcurrent device
III. at the source if the system has no disconnecting means or overcurrent device

(a) I only (b) II only (c) III only (d) I, II or III

53. Where a change occurs in the size of the ungrounded conductor ____.

(a) the neutral can be reduced two sizes
(b) the grounded conductor can be reduced 70%
(c) a similar change may be made in the size of the grounded conductor
(d) the only reduction for the neutral is household ranges

54. Which of the following applies to Class I Division I locations?

(a) ignitible flammable gases and vapors
(b) grain silos
(c) ignitible fibers or flyings
(d) combustible dust

55. Tap devices used in FC assemblies shall be rated at not less than ____ amps or more than 300 volts, and they shall be color-coded in accordance with the requirements of 322.120(C).

(a) 20 (b) 15 (c) 30 (d) 40

56. A ____ is an electrical device that converts single-phase power to three-phase electrical power.

(a) rectifier (b) generator (c) phase converter (d) phase splitter

57. Underground service conductors must have a rating not smaller than ____.

(a) #3 (b) #4 (c) #6 (d) #8

58. The Code provides that in Class I, Division I locations conduit seals shall be placed not farther from spark-producing devices than ____ inches.

(a) 18 (b) 24 (c) 12 (d) 30

59. Where conduit is threaded in the field, it is assumed that a standard cutting die should provide ____ taper per foot.

(a) 1/4" (b) 1/2" (c) 3/4" (d) 1'

60. How would you connect the grounded system conductor to the grounding electrode?

(a) grounded conductor **(b) grounding conductor**
(c) bonding jumper **(d) bonding jumper main**

61. The height of a circuit breaker used as a switch shall not exceed ____ feet above the floor.

(a) 4 (b) 4 1/2 (c) 5 (d) 6' 7"

62. Equipment intended to interrupt current at other than ____ shall have an interrupting rating at nominal circuit voltage sufficient for the current that must be interrupted.

(a) fault levels (b) maximum current (c) peak current (d) short circuit levels

63. Ground-fault protection that functions to open the service disconnecting means ____ protect(s) service conductors or the service disconnecting means.

(a) will (b) will not (c) adequately (d) totally

64. Enclosures for isolating switches in Class I, Division II locations ____.

(a) may be of general-purpose type
(b) must be explosion-proof type
(c) must have interlocking devices
(d) may not have doors in hazardous area

65. Which of the following statements about the protection of NM sheathed cable from physical damage is (are) correct?

I. When passing through a floor the cable shall be enclosed in pipe or conduit extending at least 6 inches above the floor.
II. When run across the top of the floor joists in an accessible attic, the cable shall be protected by guard strips.

(a) I only (b) II only (c) both I and II (d) neither I nor II

66. The type of grounding receptacle or plug required in the hazardous location of an anesthetizing location shall be listed as ____.

(a) special grade receptacle
(b) Class I, Division I type receptacle
(c) Class II, Division II receptacle
(d) Class I, Group C type receptacle

67. Lengths of not more than ____ of AC cable at terminals where flexibility is necessary does not have to be supported.

(a) 28" (b) 2' (c) 30" (d) 3'

68. Any room or location in which flammable anesthetics are stored shall be considered to be a ____ location from floor to ceiling.

(a) Class I, Division I **(b) Class I, Division II**
(c) Class II, Division II **(d) nonhazardous**

69. Circuits containing neon tubing installations exclusively shall not be rated in excess of ____ amps.

(a) 20 (b) 30 (c) 40 (d) 50

70. Where raceway-type service masts are used, all raceway fittings shall be ____ for use with service masts.

(a) identified (b) approved (c) heavy-duty (d) none of these

2005

OPEN BOOK EXAM #12

70 QUESTIONS
TIME LIMIT - 2 HOURS

TIME SPENT ______ **MINUTES**

SCORE ______ **%**

MASTER OPEN BOOK EXAM #12 Two Hour Time Limit

1. The equipment grounding conductor in type NM cable for 15, 20 and 30 ampere branch circuits ____.

(a) may be at least one size smaller than the insulated circuit conductor
(b) must be the same size as the insulated circuit conductors
(c) is required only with aluminum cable
(d) none of these

2. Thermoplastic insulation may stiffen at temperatures colder than minus ____ degrees C requiring care be exercised during installation at such temperatures.

(a) 5 (b) 10 (c) 15 (d) 30

3. Service heads for service conductors shall be ____.

(a) raintight (b) weatherproof (c) rainproof (d) watertight

4. Buildings of multiple occupancy shall be permitted to have ____ separate sets of service entrance conductors which are tapped from one service drop.

(a) one (b) two (c) two or more (d) no

5. The clearance for 6900 volt to ground conductors over open land for grazing shall not be less than ____.

(a) 12' (b) 16' (c) 18.5' (d) 20'

6. At what angle does a header attach to the cells?

(a) parallel (b) straight (c) right angle (d) none of these

7. In Class I, Division I locations, the Code requires conduit seals adjacent to boxes containing splices if the conduit is equal to or larger than ____.

(a) 3/4" (b) 1 1/2" (c) 1" (d) 2"

8. Every electric sign of any type, fixed or portable, shall be ____.

(a) listed (b) approved (c) permanently wired (d) electrically isolated

9. In motion-picture studios, feeder conductors to the stage may be protected, with respect to ampacity, at a maximum value of ____ percent.

(a) 200 (b) 250 (c) 400 (d) 500

10. Service conductors run above the top level of a window shall be permitted to be less than the ____ requirement.

(a) 3' (b) 6' (c) 8' (d) 10'

11. Cells in jars of conductive material shall be installed in trays of nonconductive material with not more than ____ cells in the series circuit.

(a) 16 (b) 18 (c) 20 (d) 24

12. Signs operated by electronic or electromechanical controllers located external to the sign shall have a disconnecting means located ____.

(a) within sight of sign
(b) within sight from controller location
(c) only in the controller
(d) only external to the controller

13. Metallic enclosures of reactors and adjacent metal parts shall be installed so that the ____ from induced circulating currents will not be hazardous to personnel or constitute a fire hazard.

(a) heat (b) arc (c) temperature rise (d) fumes

14. What is the service conductor demand load for the following 20 outlets at a marina?

17 - 30 amp receptacles 3 - 50 amp receptacles

(a) 660 amps (b) 528 amps (c) 462 amps (d) 396 amps

15. A 240 volt single-phase room air conditioner shall be considered as a single motor unit if its rating is not more than ____ amps.

(a) 20 (b) 30 (c) 40 (d) 50

16. Listed or labeled equipment shall be installed, used, or both, in accordance with ___ .

(a) the job specifications
(b) the plans
(c) the instructions given by the authority having jurisdiction
(d) the instructions included in the listing or labeling

17. Which of the following statements about the connection of an small appliance receptacle outlet at a dwelling is (are) correct?

I. The refrigerator can be plugged into it.
II. The outdoor receptacle outlet may be connected to one of the required small appliance circuits.

(a) I only (b) II only (c) both I and II (d) neither I nor II

18. Silicone rubber insulated fixture wire SF-1 should be limited to use where the voltage does not exceed ____ volts.

(a) 500 (b) 300 (c) 200 (d) 100

19. On circuits of 600 volts or less, overhead spans up to 50 feet in length shall have conductors not smaller than ____.

(a) #14 (b) #12 (c) #6 (d) #10

20. A pool recirculating pump motor receptacle shall be permitted not less than ____ feet from the inside walls of the pool.

(a) 5 (b) 8 (c) 10 (d) 15

21. A bare #4 conductor may be concrete-encased and serve as the grounding electrode when at least ____ feet in length.

(a) 25 (b) 15 (c) 10 (d) 20

22. For limited flexibility for motor connections in a Class I, Division II location, flexible conduit ____.

(a) must be explosion-proof
(b) must be liquid-tight flexible conduit or equal
(c) may be standard flexible metal conduit
(d) shall not be used

23. Flexible metal conduit shall be secured at which of the following?

(a) at intervals not exceeding 4 1/2 feet
(b) within 12 inches on each side of a box where fished
(c) where fished
(d) lengths not exceeding 3' at motor terminals

24. The ampacity of capacitor circuit conductors shall not be less than ____ percent of the rated current of the capacitor.

(a) 100 (b) 125 (c) 135 (d) 150

25. Cables and receptacles associated with the information technology equipment shall be permitted under a raised floor, provided ____.

I. openings in raised floors for cables protect cables from abrasions
II. ventilation in the underfloor area is used only for information technology equipment
III. the raised floor is of suitable construction and the underfloor area is accessible

(a) I only (b) II only (c) III only (d) I, II and III

26. When bare grounded conductors are used with insulated conductors, their ampacities are limited to ____.

(a) 60 degrees C
(b) 75 degrees C
(c) 90 degrees C
(d) that permitted for the adjacent insulated conductors

27. Fixtures shall be so constructed or installed that adjacent combustible material will not be subjected to temperature in excess of ____ degrees C.

(a) 75 (b) 90 (c) 185 (d) 140

28. Metal enclosures for grounding electrode conductors shall be ____.

(a) rigid conduit only
(b) not less than 3/4" in diameter
(c) bonded
(d) electrically continuous

29. Storerooms and similar areas adjacent to aircraft hangars but effectively isolated shall be designated ____.

(a) Class I, Division II
(b) Class II, Division I
(c) Class II, Division II
(d) shall not be classified

30. With consideration to mobile homes, which of the following major appliances, other than built-in, are NOT considered portable if cord connected?

(a) refrigerators
(b) range equipment
(c) clothes washers
(d) water heater

31. Power feed, grounding connection, and shield system connection between the FCC system and other wiring systems shall be accomplished in a ____.

(a) transition assembly (b) raceway (c) trench (d) none of these

32. Audible and visual signal devices shall be provided, where practicable to indicate ____.

I. derangement of emergency source
II. that the battery charger is not functioning
III. that the battery is carrying load

(a) I only (b) II and III only (c) I and II only (d) I, II and III

33. Type MC cable installed outside of buildings or as aerial cable shall comply with NEC Article ____.

(a) 225 (b) 710.3 (c) 300.5 (d) 225 and 396

34. A ground ring encircling the building, in direct contact with the earth at a depth below earth surface shall not be less than ____ feet, consisting of at least 20' of bare copper not smaller than #2.

(a) 2 (b) 2 1/2 (c) 3 (d) 4

35. Each patient bed in a critical care area shall be provided with a minimum of ____ receptacle(s).

(a) one duplex (b) single (c) two duplex (d) six

36. Conductors supplying several motors shall have an ampacity equal to the sum of the full-load current rating of all the motors plus ____% of the highest rated motor in the group.

(a) 25 (b) 80 (c) 100 (d) 125

37. Rigid conduit shall be ____ every 10 feet as required by section 110.21.

(a) stamped (b) clearly and durably identified
(c) marked (d) none of these

38. The grounding electrode conductor shall be ____ and shall be installed in one continuous length without a splice or joint.

I. solid II. solid or stranded III. insulated, covered or bare

(a) I only (b) I and III (c) II and III (d) III only

39. Branch circuits supplying two or more outlets for fixed space heating equipment in a dwelling shall be rated at ____ amperes.

(a) 15-20-25-30 (b) 15-20-30-40 (c) 15-20-30 (d) 20-30-40

40. Field bends or modifications shall be so made that the ____ of the cable tray system and support for the cables shall be maintained.

(a) temperature (b) electrical continuity (c) strength (d) rigidity

41. Protection shall be provided for exposed conductors and equipment during process of manufacturing, ___ at the building site.

I. erection II. in transit III. packaging

(a) I only (b) II only (c) III only (d) I, II and III

42. The words "thermally protected" appearing on the nameplate of a motor or motor compressor indicate that the motor is provided with a ____.

(a) fuse (b) breaker (c) thermal protector (d) switch

43. Hanging fixtures where located directly above any part of the bathtub shall be so installed that the fixture is not less than ____ feet above the top of the bathtub.

(a) 4 (b) 6 (c) 8 (d) none of these

44. Conductor overload protection is not required if ____.

(a) conductors are oversized by 125%
(b) conductors are part of a limited-energy circuit
(c) interruption of the circuit can create a hazard
(d) none of the above

45. The largest stranded conductor permitted to be connected to terminals by means of upturned lugs is ____ AWG.

(a) #8 (b) #6 (c) #10 (d) #12

46. Electrical installations in hollow spaces, vertical shafts, and ventilation or air-handling ducts shall be so made that the possible spread of fire or products of combustion will not be ____.

(a) substantially increased (b) allowed (c) exposed (d) underrated

47. A branch circuit feeding a sign which has a combination of lamps and transformers shall not exceed the rating of ____ amps.

(a) 15 (b) 20 (c) 30 (d) 50

48. The minimum length of free conductor left at each outlet and switch point in a dwelling shall not be less than ____ inches.

(a) 4 (b) 6 (c) 8 (d) 10

49. Knob and tube wiring splices shall be ____ unless approved devices are used.

(a) taped (b) bolted (c) clamped (d) soldered

50. Each commercial building and each commercial occupancy accessible to pedestrians shall have at least one outside sign outlet branch circuit rated at ____ amps.

(a) 15 (b) 20 (c) either (a) or (b) (d) neither (a) nor (b)

51. Open individual service conductors in wet locations, maximum 600 volts, supported every 9 feet should have a minimum clearance between conductors of ____.

(a) 2 1/2" (b) 3" (c) 6" (d) 12"

52. Angle-pull dimensional requirements apply to junction boxes only when the size of conductor is equal to or larger than ____.

(a) #0 (b) #4 (c) #3/0 (d) #6

53. According to the Code, conductors on poles, where not placed on racks or brackets, shall be supported not less than ____.

(a) 6" (b) 12" (c) 18" (d) 24"

54. Wood braces used in structural mounting of boxes shall have a cross-section not less than nominal ____.

(a) 3/4" x 1 1/2" (b) 1" x 1 1/2" (c) 1" x 2" (d) 3/4" x 2"

55. Pull-type canopy switches shall not be located more than ____ inches from the center of the canopy.

(a) 1 1/2" (b) 2" (c) 3" (d) 3 1/2"

56. Plug fuses of the Edison-base type shall be used ____.

(a) where overfusing is necessary
(b) only as replacement items in existing installations
(c) as a replacement for type S fuses
(d) only for 50 amps and above

57. Where motors are provided with terminal housing, the housing shall be ____ and of substantial construction.

(a) plastic (b) metal (c) either (a) or (b) (d) neither (a) nor (b)

58. A storage battery supplying emergency lighting and power shall maintain not less than 87 1/2 percent of full voltage at total load for a period of at least ____.

(a) 2 hours (b) 1 1/2 hours (c) 1 hour (d) 1/4 hour

59. The internal depth of outlet boxes intended to enclose flush devices shall be at least ____.

(a) 1/2" (b) 7/8" (c) 15/16" (d) 1"

60. The 150 va rating per 2 feet of lighting track ____.

I. will limit the number of fixtures allowed
II. is for load calculations only
III. it limits the length of track that can be run

(a) I only (b) II and III only (c) II only (d) I, II and III

61. When supplying nominal 120 volt rated room air-conditioner, the length of the flexible supply cord shall not exceed ____ feet.

(a) 4 (b) 6 (c) 8 (d) 10

62. Sheet metal auxiliary gutters shall be supported throughout their entire length not exceeding ____ feet.

(a) 5 (b) 6 (c) 8 (d) 10

63. Each patient bed location shall be supplied by at least ____.

(a) five receptacles
(b) one branch circuit
(c) two branch circuits
(d) three branch circuits

64. Insulating bushings are required on conduit entering boxes, gutters, etc. if it contains conductors as large as ____.

(a) #2 (b) #4 (c) #0 (d) #6

65. _____ is the wall and arch that separates the stage from the auditorium.

(a) Stage drop (b) Proscenium (c) Acrolein (d) Butadiene

66. A fixture that weighs more than 6 pounds or exceeds ____ inches in any dimension shall not be supported by the screw shell.

(a) 12 (b) 16 (c) 18 (d) 24

67. Bored holes in wood members for cable or raceway-type wiring shall be bored so that the edge of the hole is not less than ____ inches from the nearest edge.

(a) 1 1/4" (b) 1 1/8" (c) 1 1/2" (d) 1 1/16"

68. Unless specified otherwise, live parts of electrical equipment operating at ____ volts or more shall be guarded.

(a) 12 (b) 15 (c) 50 (d) 24

69. Generators operating at ____ volts or less and driven by individual motors shall be considered as protected by the overcurrent device protecting the motor if these devices will operate when the generators are delivering not more than 150% of their full load rated current.

(a) 65 (b) 70 (c) 75 (d) 85

70. Where necessary, a shallow outlet box not less than ____ inches deep may be used.

(a) 1/2 (b) 3/4 (c) 1 (d) 1 1/4

2005

OPEN BOOK EXAM #13

70 QUESTIONS
TIME LIMIT - 2 HOURS

TIME SPENT [ ] MINUTES

SCORE [ ] %

MASTER OPEN BOOK EXAM #13 Two Hour Time Limit

1. Tap conductors in a metal raceway for recessed fixture connections shall be limited to ____ feet in length.

(a) 2 (b) 4 (c) 6 (d) 10

2. The number of square feet that each made plate electrode should present to the soil is ____.

(a) 4 square feet (b) 3 square feet (c) 2 square feet (d) 1 square foot

3. An autotransformer starter shall provide ____.

I. an "off position"
II. a running position
III. at least one starting position

(a) I only (b) II only (c) I and II only (d) I, II and III

4. When service-entrance conductors exceed 1100 kcmil for copper, the bonding jumper must be at least ____ percent of the largest phase conductor.

(a) 15 (b) 10 (c) 12 1/2 (d) 25

5. Nonmetallic cabinets in a wet location shall be permitted ____ airspace between concrete, masonry tile or similar wall.

(a) without (b) with 1/4" (c) with 1/8" (d) 1/16"

6. A 30 amp remote panelboard containing conductors protected by a 30 amp fuse shall have a ____ equipment grounding conductor minimum.

(a) #12 (b) #10 (c) #8 (d) #6

7. The connection of a grounding electrode conductor to a driven ground rod is ____.

(a) required to be visible **(b) required to be accessible**
(c) required to be readily accessible **(b) not required to be accessible**

8. The maximum rating of a plug fuse is ____ amps.

(a) 20 (b) 30 (c) 15 (d) 40

9. A controller for a motor-compressor, serving more than one motor-compressor and other loads, shall have ____.

I. a continuous duty, F.L.C. rating
II. A locked-rotor current rating **not** less than the combined load as determined in accordance with section 440.12b

(a) I only (b) II only (c) either I or II (d) both I and II

10. A multi-wire branch circuit may supply ____.

I. only one utilization equipment II. ungrounded conductors that are opened simultaneously

(a) I only (b) II only (c) both I and II (d) neither I nor II

11. Electric vehicle cable type EVJ ____.

I. comes in sizes #18-#500 kcmil II. is for extra hard usuage III. has thermoset insulation

(a) I only (b) II only (c) III only (d) I, II and III

12.Flexible cords and cables shall be so connected to devices and to fittings that ____ will not be transmitted to joints or terminals.

(a) shock (b) tension (c) heat (d) voltage

13. In an industrial commercial (loft) building, where the actual number of general purpose receptacle outlets are unknown, an additional load of ____ volt-amps per square foot shall be included in the load calculation.

(a) 1/2 va (b) 1 va (c) 1 1/2 va (d) none of these

14. The branch circuit conductors to one or more units of a data processing system shall have an ampacity of what percent of the total connected load?

(a) 80 (b) 100 (c) 125 (d) 200

15. When a controller is **not** within sight from the motor location the disconnect shall be capable of being ____ in the open position.

(a) up (b) down (c) locked (d) shut-off

16. Circuits for electrical cranes operating in Class III locations over combustible fibers shall not be ____.

(a) grounded (b) pulled in raceways (c) spliced (d) THWN conductors

17. Single conductors in a cable tray shall be securely bound in circuit groups to prevent ____ due to fault-current magnetic forces unless single conductors are cabled together, such as triplexed assemblies.

(a) current unbalance **(b) inductive reactance**
(c) excessive movement **(d) voltage surges**

18. Class II hazardous location is where ____.

(a) gases and vapors are present
(b) combustible dust is present
(c) fibers and flyings are present
(d) radioactive material is present

19. A transverse metal raceway for electrical conductors, furnishing access to predetermined cells of a precast cellular concrete floor, which permits installation of conductors from a distribution center to the floor cells is called ____.

(a) an underfloor raceway **(b) a header**
(c) a cellular raceway **(d) a mandrel**

20. The service disconnecting means shall be installed at a readily accessible location either outside of a building or structure, or inside ____ of the service conductors.

(a) within 3 steps (b) within 6 feet (c) within 8 feet (d) nearest the point of entrance

21. Heaters installed within ____ feet of the outlet of an air-moving device, heat pump, A/C, elbows, baffle plates, or other obstructions in duct work may require turning vanes, pressure plates, or other devices on the inlet side of the duct heater to assure an even distribution of air over the face of the heater.

(a) 2 (b) 3 (c) 4 (d) 6

22. Which of the following statements about MI cable is correct?

(a) it may be used in any hazardous location
(b) a single run of cable shall not contain more than the equivalent of 4 quarter bends
(c) it shall be securely supported at intervals not exceeding 10 feet
(d) it may be mounted flush on supporting surfaces in a wet location

23. Receptacles located over ____ feet above the floor are not counted in the required number of receptacles along the wall.

(a) 4 (b) 5 (c) 5 1/2 (d) none of these

24. The maximum size of a receiving station outdoor antenna conductor, where the span is 75 feet, shall be at least ____ if a copper-clad steel conductor is used.

(a) #10 (b) #12 (c) #14 (d) #17

25. A main common-return conductor in the electromagnetic supply shall not be less than ____.

(a) #14 (b) #18 (c) #20 (d) #22

26. The minimum thickness of the sealing compound in Class I, Division I and II locations shall not be less than the trade size of the conduit and in no case less than ____.

(a) 3/16" (b) 3/8" (c) 1/2" (d) 5/8"

27. What is the cross-sectional area of a 2" schedule 40 PVC conduit?

(a) 3.291 (b) 1.34 (c) 2.016 (d) 2.067

28. Multi-outlet assembly may be used ____.

(a) where concealed **(b) in hoistways**
(c) in dry locations **(d) in storage battery rooms**

29. The rating of the surge arrester shall be ____ the maximum continuous phase-to-ground power frequency voltage.

I. equal to II. less than III. greater than

(a) I only (b) I or II (c) I or III (d) II only

30. Which of the following is an acceptable wiring method for the forming shell for underground sound equipment?

I. rigid metal conduit II. IMC brass III. EMT IV. rigid nonmetallic conduit

(a) I only (b) II and III only (c) I, II and IV (d) IV only

31. Hoistway is a ____ in which an elevator or dumbwaiter is designed to operate.

(a) shaftway (b) hatchway (c) well hole (d) all of these

32. Above ground storage tanks shall be classified ____ when the space between 5 feet and 10 feet from open end of vent extends in all directions.

(a) Class I, Division I
(b) Class I, Division II
(c) Class II, Division I
(d) Class II, Division II

33. Tests are to be performed and made available to the inspector on all cords sets and receptacles used for temporary wiring on construction sites. All tests shall be performed ____.

I. at intervals not exceeding 3 months
II. when there is evidence of damage
III. before use on the construction site

(a) I only (b) II only (c) III only (d) I, II and III

34. In other than dwelling-type occupancies, each electrically heated appliance or group of appliances intended to be applied to combustible material shall be provided with a ____.

(a) light (b) thermostat (c) signal (d) warning

35. The grounding conductor for secondary circuits of instrument transformers and for instrument cases shall **not** be smaller than #12 ____.

I. metal II. aluminum III. copper

(a) I only (b) II only (c) III only (d) I, II and III

36. Fixtures or lampholders should have no live parts normally exposed to contact unless they are ____.

(a) rosette type 6 feet above the floor
(b) cleat type located at least 8 feet above the floor
(c) both (a) and (b)
(d) neither (a) nor (b)

37. For a feeder supplying household cooking equipment and electric clothes dryers the minimum unbalanced load on the neutral conductor shall be considered as ____ percent of the load on the ungrounded conductors.

(a) 40 (b) 50 (c) 70 (d) 80

38. Where a double-throw knife switch has a vertical throw, a(n) ____ shall be provided to hold the blades in the open position when so set.

(a) closed device (b) automatic device (c) integral mechanical means (d) tamperproof device

39. Means shall be provided to disconnect the ____ of all fixed electric space heating equipment from all ungrounded conductors.

I. heater
II. motor
III. controller
IV. supplementary overcurrent protective devices

(a) I and II only (b) II and IV only (c) I and IV (d) I, III and IV only

40. Where required, drawings for feeder installations shall be provided prior to the ____.

(a) completion of installation **(b) installation of the feeders**
(c) removal of the feeders **(d) all of these**

41. A single electrode consisting of a ____ which does not have a resistance to ground of 25Ω or less shall be augmented by one additional electrode.

I. rod II. pipe III. plate

(a) I only (b) II only (c) III only (d) I, II or III

42. How often must 1" threaded rigid metal conduit be supported?

(a) every 10 feet (b) every 12 feet (c) every 14 feet (d) every 20 feet

43. A thermal barrier shall be required if the space between the resistors and reactors and any conductor would be ____ inches.

(a) 12 (b) 15 (c) 18 (d) 24

44. If the terminal for the equipment grounding conductor is not visible on the receptacle, the conductor entrance hole shall be marked with the ____.

(a) letter G (b) letter GR (c) word ground (d) any of these

45. Signs and ____ shall be marked with input amperes at full load and input voltage.

(a) markers (b) incandescent lighting (c) outline lighting systems (d) neon lighting

46. Which of the following is **not** required on a motor nameplate?

(a) horsepower **(b) maker's name**
(c) watts **(d) voltage**

47. Fire protective signaling circuits shall be grounded, the only exception is DC power-limited circuits having a maximum current of ____ amperes.

(a) 0.300 (b) 1.5 (c) 0.030 (d) 2.34

48. FCC cable can be installed under carpet squares no larger than ____ inches.

(a) 24 (b) 30 (c) 36 (d) 48

49. Industrial machinery is defined as ____.

I. a portable machine used to shape or form plastic
II. a power-driven machine not portable by hand

(a) I only (b) II only (c) I and II (d) neither I nor II

50. The TVSS shall be marked with a short circuit current rating and shall not be installed at a point on the system where the available fault current is in excess of that rating. This marking requirement shall not apply to ____.

(a) motors (b) controllers (c) lights (d) receptacles

51. Where reduced heating of the conductors results from motors operating on duty-cycle, intermittently, or from all motors not operating at one time, the feeder conductors ____

(a) are not allowed to have the ampacity reduced
(b) may have an ampacity less than specified if acceptable to the authority having jurisdiction
(c) must be sized no smaller than 125% of the largest motor connected to the feeder
(d) must be sized not smaller than 125% of the largest motor plus other loads

52. Splices and taps shall be permitted within a wireway provided they are accessible. The conductor including splices and taps shall not fill the wireway to more than ____ percent of its area at that point.

(a) 25 (b) 80 (c) 125 (d) 75

53. Escalator motors shall be classified as ____ duty.

(a) intermittent (b) varying (c) short-time (d) continuous

54. An approved seal shall be provided in each conduit run entering or leaving a dispenser or any cavities or enclosures in direct communication therewith. The sealing fitting shall be _____ .

(a) concrete-tight
(b) 3/4" minimum thickness
(c) the last fitting after the conduit emerges from the earth or concrete
(d) the first fitting after the conduit emerges from the earth or concrete

55. The grounding conductor for a TV antenna shall not be smaller than a ____ copper.

(a) #6 (b) #8 (c) #10 (d) #12

56. In land areas not subject to tidal fluctuation, the electrical datum plane is a horizontal plane ____ above the highest water level for the area occurring under normal circumstances.

(a) 2' (b) 4' (c) 6' (d) 8'

57. What is the area of square inch for a #14 RHH with an outer covering?

(a) .0209 (b) .0293 (c) .0135 (d) .0206

58. The maximum length of a bonding jumper on the outside of a raceway is ____ feet.

(a) 3 (b) 6 (c) 25 (d) none of these

59. A grounding electrode connection that is encased in concrete or directly buried shall ___.

(a) be made accessible **(b) be made only by exothermic welding**
(c) be a minimum #4 bare **(d) not be required to be accessible**

60. What is the minimum size fixture wire?

(a) #16 (b) #18 (c) #20 (d) #22

61. Where practicable, dissimilar metals in contact anywhere in the system shall be avoided to eliminate the possibility of ____.

(a) hysteresis (b) galvanic action (c) specific gravity (d) resistance

62. Types of resistive heaters are ____.

I. heating blanket II. heating tape III. heating barrel

(a) I only (b) II only (c) I and II only (d) I, II and III

63. Portable outdoor signs shall be protected by ____.

(a) current limiting fuses **(b) instantaneous breakers**
(c) GFCI **(d) time delay fuses**

64. Which of the following need not be grounded?

(a) motion picture projection equipment **(b) electrically operated cranes**
(c) metal service raceways **(d) electric furnace (industrial)**

65. In auxiliary gutters, the minimum clearance between bare current-carrying metal parts of different potential mounted on the same surface will not be less than ____ for parts that are held free in the air.

(a) 3/4" (b) 1" (c) 1 1/2" (d) 2"

66. A forming shell shall be provided with a number of grounding terminals that shall be ____ the number of conduit entries.

(a) one more (b) two more (c) same as (d) none of these

67. Suitable covers shall be installed on all boxes, fittings, and similar enclosures to prevent accidental contact with ____ parts or physical damage to parts or insulation. Over 600 volts nominal.

(a) energized (b) mechanical (c) electrical (d) none of these

68. A 1000 watt incandescent lamp shall have a ____ base.

(a) mogul (b) standard (c) admedium (d) copper

69. What kind of lighting loads does the Code state there shall be no reduction in the size of the neutral conductor?

(a) dwelling unit (b) hospital (c) nonlinear loads (d) motel

70. The volume of a wiring enclosure (box) shall be the total volume of the ____, and, where used, the space provided by plaster rings, domed covers, extension rings, etc., that are marked with their volume in cubic inches, or are from boxes the dimensions of which are listed in Table 314.16(A).

(a) enclosure (b) outlet (c) assembled sections (d) none of these

2005

OPEN BOOK EXAM #14

70 QUESTIONS
TIME LIMIT - 2 HOURS

TIME SPENT ______ **MINUTES**

SCORE ______ **%**

MASTER OPEN BOOK EXAM #14 **Two Hour Time Limit**

1. Shore-power receptacles rated not less than 30 amperes or more than 50 amperes shall be ____.

(a) GFCI
(b) of the locking and grounding type
(c) tamperproof
(d) able to be locked in the "off" position

2. Which of the following pool parts are required to be bonded together?

I. All fixed metal parts within 5 feet of the inside walls.
II. All forming shells and mounting brackets of a no-niche fixture unless listed for a low voltage system.
III. All metal parts of an underwater sound system.

(a) I only (b) II only (c) I and II only (d) I, II and III

3. When installing office furnishings, receptacle outlets, ____ be located in lighting accessories.

(a) single-type only can (b) duplex-type only can (c) shall (d) shall not

4. Elevators shall have a single means for disconnecting all ungrounded main power supply conductors for each unit; ____.

(a) this does not include the emergency power service if the system is automatic
(b) this does include the emergency power service
(c) this does not include the emergency power service
(d) no elevators are to operate on emergency power systems

5. A motor operating at 480 volts, the minimum line terminal housing spacing between fixed line terminals must be ____ inch.

(a) 1/4 (b) 1/2 (c) 3/8 (d) 5/8

6. What is the minimum thickness for a 6" x 4" x 3 1/4" box?

(a) .0625" (b) .0747" (c) 15 MSG (d) 16 MSG

7. Capacitors containing more than ____ gallons of flammable liquid shall be enclosed in vaults or outdoor fenced enclosures.

(a) 3 (b) 5 (c) 7 (d) 10

8. The branch circuit overcurrent devices in emergency circuits shall be ____.

(a) of the reset type only **(b) a slow-blow type**
(c) accessible to only authorized personnel **(d) painted yellow**

9. An electrically operated organ shall have both the generator and motor frames grounded or ____.

(a) the generator and motor shall be effectively insulated from ground and from each other
(b) the generator and motor shall be effectively insulated from ground
(c) the generator shall be effectively insulated from ground and from the motor driving it
(d) both the generator and motor shall have double insulation

10. Cable trays include fittings or other suitable means for ____.

I. temperature
II. electric continuity
III. changes in direction and elevation of runs

(a) I only (b) I and II only (c) III only (d) I and III only

11. A raceway contains 45 current-carrying conductors. The ampacity of each conductor shall be reduced ____ percent.

(a) 80 (b) 70 (c) 60 (d) 35

12. Multiconductor portable cables used to connect mobile equipment and machinery above 2000 volts, the conductors shall be ____.

(a) without ground (b) unshielded (c) shielded (d) MK braid

13. A pool panelboard, not part of the service equipment, shall have a grounding conductor installed between ____.

(a) its grounding terminal and a separate ground
(b) its grounding terminal and a ground rod
(c) its grounding terminal and the grounding terminal of the service equipment
(d) its grounding terminal and bonding grid

14. In communication circuits the bonding together of all separate electrodes ____.

(a) shall not be permitted
(b) shall be permitted with a minimum size jumper #4
(c) shall be permitted with a minimum size jumper #6
(d) shall be permitted with a minimum size jumper #8

15. Open motors with commutators shall be located so sparks cannot reach adjacent combustible material, but this ____.

(a) is only required for over 600 volt motors
(b) shall not prohibit these motors on wooden floors
(c) does not prohibit these motors from a Class I location
(d) none of these

16.____ shall be controlled by an externally operable switch or breaker which will open all ungrounded conductors.

I. Outline lighting II. signs III. Portable signs

(a) I only (b) II only (c) III only (d) I and II only

17. A public address system ____.

(a) is not covered in the Code **(b) has its own Code section**
(c) is covered in the Code under Art. 640 **(d) none of these**

18. 3" rigid nonmetallic conduit has a maximum spacing between supports of ____ feet.

(a) 3 (b) 5 (c) 6 (d) 8

19. Conductors which supply one or more AC transformers or DC rectifier arc welders shall be protected by an overcurrent device rated or set at not more than ____ percent of the conductor rating.

(a) 70 (b) 80 (c) 125 (d) 200

20. The battery voltage computed on the basis of ____ volts per cell for the lead-acid type and ____ volts per cell for the alkali type.

(a) 1.5 for lead-acid, 2.0 for the alkali
(b) 2.0 for lead-acid, 1.5 for the alkali
(c) 2.0 for lead-acid, 1.2 for the alkali
(d) 1.2 for lead-acid, 2.0 for the alkali

21. Optional standby systems are typically installed to provide an alternate source of electric power for such loads as ____ systems.

I. communications II. data processing III. refrigeration

(a) III only (b) I and III only (c) II and III only (d) I, II and III

22. Feeder circuits supplying power to irrigation machines, shall have an equipment grounding conductor _____.

(a) not smaller than #6 cu
(b) sized according to Table 250.66
(c) sized according to Table 250.122
(d) not smaller than #8 cu

23. Which of the following is **not** a true statement concerning an equipment grounding conductor?

(a) Under certain conditions, equipment grounding conductors may be required to be larger than circuit conductors.
(b) Under certain conditions, equipment grounding conductors may be run in parallel.
(c) One size of equipment grounding conductors shall be increased to compensate for voltage drop.
(d) One equipment grounding conductor may serve multiple circuits.

24. In a recreational vehicle, which major appliance, other than built-in does the Code consider portable if cord-connected?

I. refrigerator II. clothes washers III. gas range equipment

(a) I only (b) I and II only (c) II only (d) I, II and III

25. A copper bus bar is 4" wide by 1/2" thick. What is the ampacity?

(a) 500 amps (b) 1000 amps (c) 1500 amps (d) 2000 amps

26. Where rigid PVC conduit is used as a raceway system in bulk storage plant wiring, the raceway shall include ____.

(a) sunlight resistant listing
(b) an equipment grounding conductor
(c) a bushing with double locknuts
(d) PVC raceway is not permitted

27. All devices excluding lighting and appliance branch circuit panelboards provided with terminals for the attachment of conductors and intended for connection to more than one side of the circuit shall have ____ properly marked for identification.

(a) conductors (b) terminals (c) sides (d) none of these

28. There shall be an air space of at least ____ between walls, back, gutter partition, if of metal, or door of any cabinet, or cut out box and nearest exposed current-carrying parts of devices mounted within the cabinet where the voltage exceeds **251** volts.

(a) 1/4" (b) 1/2" (c) 1" (d) 1 1/2"

29. For equipment rated 1200 amperes or more 600 volts or less, and over 6 feet wide, containing overcurrent devices, switching devices, or control devices, there shall be one entrance not less than ___ inches wide and 6 1/2 feet high at each end.

(a) 24 (b) 30 (c) 36 (d) 48

30. A ____ is a tank or vat in which electrochemical reactions are caused by applying electrical energy for the purpose of refining or producing usable materials.

(a) electrolytic cell (b) heating boiler (c) radiant heat (d) duct heater

31. Underground service conductors carried up a pole must be protected from mechanical injury to a height of at least ____ feet.

(a) 10 (b) 8 (c) 12 (d) 15

32. All operating control and signal circuits used in elevators, dumbwaiters and escalators shall be wired with ____ wire or larger.

(a) #14 (b) #16 (c) #18 (d) #20

33. A crane rail when used as a conductor shall be ____.

I. grounded by fittings used for suspension or attachment of the rail
II. grounded by a bonding conductor to a water pipe
III. effectively grounded at the transformer

(a) I only (b) I and II only (c) I and III only (d) II and III only

34. Threaded boxes not over 100 cubic inches, that contain devices, shall be considered to be adequately supported if two or more circuits are threaded into the box wrenchtight and on the same side and if each conduit is supported within ____ inches of the box.

(a) 36 (b) 30 (c) 24 (d) 18

35. All lights and any receptacles adjacent to the mirror(s) and above the dressing table counters in dressing rooms of theaters shall be controlled by wall switches installed in the ____.

(a) dressing rooms (b) control room (c) projection room (d) stage office

36. Class 1 circuits and power supply circuits shall be permitted to occupy the same raceway only where the equipment powered is ____.

(a) low voltage (b) a fire alarm system (c) functionally associated (d) AC/DC

37. Induction generating equipment on systems with significant ____ may become self-excited upon loss of primary source and experience severe over-voltage as a result.

(a) voltage (b) amperage (c) induction (d) capacitance

38. In general, Class II control circuits and power circuits ____.

I. may occupy the same raceway II. shall be installed in different raceways

(a) I only (b) II only (c) I and II (d) none of these

39. A boiler employing resistance-type heating elements rated more than 48 amperes and not contained in an ASME rated and stamped vessel shall have the heating elements subdivided into loads not exceeding ____ and protected at not more than ____ amperes.

(a) 24...48 (b) 48...60 (c) 24...60 (d) 48...80

40. Electrical nonmetallic tubing is permitted to be used in sizes up to ____.

(a) 1" (b) 2" (c) 3" (d) 4"

41. In agricultural buildings all cables shall be secured within ____ inches of a box.

(a) 6 (b) 8 (c) 12 (d) 18

42. Because aluminum is not a magnetic metal, there will be ____ present when aluminum conductors are grouped in a wireway.

(a) no heat due to voltage (b) no heating due to hysteresis
(c) no induced currents (d) none of these

43. A residence has a front entrance on the north side of the house along with an attached garage with an 8' wide door, also a back entrance door on the south side of the house. How many lighting outlets are required for these outdoor entrances?

(a) 1 (b) 2 (c) 3 (d) none of these

44. Heating panels or heating panel sets in poured concrete shall not exceed 33 watts per square foot of heated area or ____ watts per linear foot of cable.

(a) 10 1/2 (b) 16 1/2 (c) 25 (d) 33

45. Each switchboard, switchboard section, or panelboard, if used as service equipment, shall be provided with ____.

(a) a main bonding jumper **(b) a power circuit**
(c) a battery charging panel **(d) a 4-wire delta connected system**

46. When a diesel engine is used as the prime mover of a generator to supply emergency power, how much of site fuel is required?

(a) one-half hour of fuel supply **(b) one hour of fuel supply**
(c) two hours of fuel supply **(d) three hours of fuel supply**

47. The length of a type S cord connecting a trash compactor must not exceed ____.

(a) 18" (b) 4' (c) 36" (d) 2'

48. On a 4-wire delta-connected secondary where the midpoint of one phase is grounded to supply lighting and similar loads, the phase conductor having the higher voltage to ground shall be identified by ____.

I. tagging
II. an outer finish that is orange in color
III. other effective means

(a) I only (b) II only (c) III only (d) I, II and III

49. Splices and taps shall not be located within fixture ____.

(a) splice boxes (b) arms and stems (c) pancake boxes (d) none of these

50. In normal practice, the NEC discourages the practice of switching the grounded conductor. However, in one of the following cases, the Code makes an exception and actually requires that this be adopted. Select that case.

(a) electrical circuits that lead or pass through wall mounted fixtures mounted at a height of less than 5' 6" and located in pediatric wards
(b) electrical circuits that lead into or pass through wet niche light fixtures located in swimming pools
(c) electrical circuits that lead into or pass through a gasoline dispensing pump
(d) electrical circuits that lead into or pass through paint spray booths

51. Where practicable, a separation of at least ____ feet shall be maintained between any coaxial cable and lightning conductors.

(a) 2 (b) 2 1/2 (c) 4 (d) 6

52. The neutral of a solidly grounded neutral system shall be permitted to be grounded at more than one point for ____.

I. services
II. direct buried portions of feeders employing a bare copper neutral
III. overhead portion installed outdoors

(a) I only (b) II only (c) III only (d) I, II and III

53. Where extensive metal in or on buildings may become energized and is subject to personal contact ____ will provide additional safety.

(a) adequate bonding and grounding (b) bonding
(c) suitable ground detectors (d) none of these

54. Color braid of flexible cords used to identify the use of the grounded conductor shall be finished to show a ____ color, and the braid on the other conductor or conductors finished to show a readily distinguishable solid color or colors.

I. white II. green III. natural gray IV. light blue

(a) I and II only (b) II and IV only (c) I and III only (d) III only

55. Switches, flashers, and similar devices controlling transformers shall be either rated for controlling inductive load(s) or have an ampere rating not less than ____ the ampere rating of the transformer.

(a) 100% (b) 125% (c) 200% (d) 300%

56. Solid dielectric insulated conductors operated above 2000 volts in permanent installations shall have ozone-resistant insulation and shall be ____.

(a) covered (b) protected (c) shielded (d) surface mounted

57. The ampacity requirements of x-ray equipment shall be based on ____ percent of the momentary rating of the equipment.

(a) 40 (b) 50 (c) 70 (d) 80

58. A grounding electrode conductor shall not be required for a system that supplies a ____ circuit and is derived from a transformer rated not more than 1000va.

(a) Class I (b) Class II (c) Class III (d) all of these

59. Motion picture projectors are ____.

(a) covered under theater and similar locations in the Code
(b) not covered in the Code
(c) covered in their own section of the Code
(d) part of the section in the Code on motion picture studios

60. DC conductors used for electroplating shall be protected from overcurrent by ____.

I. a current sensing device which operates a disconnecting means
II. fuses or circuit breakers
III. other approved means

(a) II only (b) I and II only (c) II and III only (d) I, II and III

61. The paralleling efficiency of ground rods longer than ____ feet is improved by spacing greater than 6 feet.

(a) 8 (b) 10 (c) 15 (d) 20

62. Equipment having an open-circuit voltage exceeding ____ volts shall not be installed in dwelling occupancies.

(a) 1000 (b) 460 (c) 600 (d) 208

63.Two-wire DC circuits used in DC system grounding in an integrated electrical system shall be permitted to be ____.

(a) ungrounded (b) uninsulated (c) over 600v (d) none of these

64. Type USE service entrance cable, identified for underground use in a cabled assembly, may have a ____ concentric.

(a) bare copper **(b) covered metal**
(c) bare aluminum **(d) covered**

65. Each fixture of each secondary circuit of tubing for electric-discharge lighting system, having an open-circuit voltage of 1500 volts, shall have clearly legible marking reading ____.

I. "Caution 1500 volts" II. "Caution High Voltage" III. "Danger High Voltage"

(a) I only (b) II only (c) III only (d) I, II and III

66. A ____ is a protective device for limiting surge voltages by discharging or bypassing surge current, and it also prevents continued flow of follow current while remaining capable of repeating these functions.

(a) surge arrester (b) auto fuse (c) fuse (d) circuit breaker

67. Motor branch-circuit, short-circuit and ground fault protection and motor overload protection shall be permitted to be combined in a single protection device where the rating or setting of the device provides the ____ protection specified in section 430-32.

(a) combined overload (b) overcurrent (c) overload (d) branch-circuit

68. The maximum ampacity for a single #500 kcmil conductor in a type IGS-EC cable shall not exceed ____ amperes.

(a) 119 (b) 168 (c) 206 (d) 238

69. For cord and attachment, plug-cord connected motor-compressor and equipment on 15 or 20 ampere branch-circuits, the rating of the attachment plug and receptacle shall **not** exceed 20 amperes at ____ volts nor 15 amperes at ____ volts.

(a) 250...125 (b) 125...250 (c) 250...250 (d) 125...125

70. Loop wiring for underfloor raceways, shall not be considered ____.

I. a splice II. a tap

(a) I only (b) II only (c) both I and II (d) neither I or II

2005

OPEN BOOK EXAM #15

70 QUESTIONS
TIME LIMIT - 2 HOURS

TIME SPENT ______ MINUTES

SCORE ______ %

MASTER OPEN BOOK EXAM #15 Two Hour Time Limit

1. In commercial garages, generally the floor area to a height of 18" above grade is designated as ____.

(a) Class I, Division I **(b) Class I, Division II**
(c) Class II, Division II **(d) Class II, Division I**

2. In using multiple grounding electrodes, they shall be separated one from the other at ____ feet distance apart.

(a) 6 (b) 8 (c) 10 (d) 12

3. The disconnecting means for a 50 hp three-phase 460v induction motor shall have an ampere rating of at least ____ amps.

(a) 126 (b) 75 (c) 91 (d) 63

4. In combustible walls or ceilings, the front edge of an outlet box or fitting may set back of the finished surface ____.

(a) 1/4" (b) 1/8" (c) 1/2" (d) not at all

5. Overcurrent protection for electric organ circuits shall not exceed ____ amps.

(a) 15 (b) 20 (c) 25 (d) none of these

6. Each transformer shall be provided with a nameplate giving the name of the manufacturer, rated kva, ____ if 25 kva and larger.

I. amount and kind of insulating liquid where used
II. frequency
III. impedance
IV. required clearances for ventilating openings

(a) I only (b) II only (c) IV only (d) I, II, III and IV

7. A single receptacle installed on an individual branch circuit shall have a rating not less than ____ percent of the rating of the branch circuit.

(a) 50 (b) 80 (c) 100 (d) 125

8. Color coding shall be permitted to identify ____ conductors where they are colored light blue and where no other conductors colored light blue are used.

(a) fire alarm (b) elevator (c) intrinsically safe (d) electrolytic cell

9. A surge arrester is a protective device for limiting surge voltages by ____ or bypassing surge current.

(a) decreasing (b) discharging (c) limiting (d) derating

10. Conductors to the hoistway door interlocks from the hoistway riser shall be ____.

(a) flame retardant **(b) type SF or equivalent**
(c) rated 200 degrees C **(d) all of these**

11. Flexible metal conduit used as a fixture whip to bond a fixture enclosure to a circuit junction box must not exceed ____ feet in length.

(a) 4 (b) 8 (c) 6 (d) 10

12. A disconnecting means serving a hermetic refrigerant motor compressor selected on the basis of the nameplate rated load current or branch circuit selection current, whichever is greater shall have an ampere rating of ____ percent of the nameplate rated load current or branch circuit selection current.

(a) 125 (b) 80 (c) 100 (d) 115

13. What size grounding conductor is required for a two-wire DC generator used in conjunction with balancer sets to obtain neutrals for a 3-wire system equipped with overcurrent devices that will disconnect the 3-wire system in case of excessive unbalancing of voltages or currents?

(a) #10 solid copper minimum size
(b) shall not be smaller than any system conductor
(c) shall not be smaller than the neutral conductor
(d) sized from Table 250.66

14. The hazardous area in a pit of a spray operation without proper vapor stop is classified as a ____ location.

(a) Class I, Division I
(b) Class I, Division II
(c) Class II, Division I
(d) Class III, Division I

15. Temporary electrical power and lighting installations shall be permitted for a period not to exceed ____ days.

(a) 90 (b) 60 (c) 30 (d) 15

16. When an outlet from an underfloor raceway is discontinued, the circuit conductors supplying the outlet ____.

(a) may be spliced
(b) may be reinsulated
(c) may be handled like abandoned outlets on loop wiring
(d) shall be removed from the raceway

17. All 125v, single-phase receptacles within ____ measured horizontally of the inside walls of a hydromassage tub shall be GFCI protected.

(a) 5' (b) 6' (c) 8' (d) 10'

18. Fixtures shall be wired with conductors having insulation suitable for the environment conditions and ____ to which the conductors will be subjected.

(a) temperature (b) voltage (c) current (d) all of these

19. Signs in wet locations shall be weatherproof and ____.

I. have drain holes positioned so there is no external obstructions
II. have at least one drain hole in every low point
III. drain holes shall not be smaller than 1/4"

(a) I and II only (b) II and III only (c) I only (d) I, II and III

20. Solid dielectric insulated conductors operated above 2000 volts in permanent installations shall have ____ insulation and shall be shielded.

(a) ozone-resistant b asbestos (c) hi-temperature (d) perfluoro-alkoxy

21.The grounding conductor shall be identified by ____.

(a) one continuous green color **(b) being bare**
(c) a continuous green color with yellow stripes **(d) any of these**

22. In general, the voltage limitation between conductors in a surface metal raceway is ____ volts.

(a) 300 (b) 600 (c) 900 (d) 1000

23. Voltage markings on cables may be misinterpreted to suggest that the cables may be suitable for ____ applications.

I. power II. electric light III. Class I

(a) I only (b) II only (c) III only (d) I, II and III

24. Locations where ignitible fibers are stored are designated as ____.

(a) Class II, Division II
(b) Class III, Division I
(c) Class III, Division II
(d) non-hazardous

25. A switch or circuit breaker should disconnect the grounded conductors of a circuit ____.

(a) by hand levers only
(b) simultaneously as it disconnects the ungrounded conductors
(c) before it disconnects the ungrounded conductors
(d) in none of the above ways

26. The space measured horizontally above a show window must have at least one receptacle for each ____ linear feet.

(a) 8 (b) 10 (c) 12 (d) 6

27. Wires run above heated ceilings and within thermal insulation, shall be derated on the basis of ____ degrees C.

(a) 86 (b) 30 (c) 50 (d) 20

28. If the allowable current carrying capacity of a conductor does not correspond to the rating of a standard size overcurrent device, the next larger size may be used provided the current does not exceed ____ amps.

(a) 200 (b) 500 (c) 800 (d) 1000

29. In a dwelling the Code requires a minimum of ____.

I. two 20 amp circuits for the small appliance circuits
II. one 20 amp circuit for the washing machine
III. one 20 amp circuit for the bathroom

(a) I only (b) I and II only (c) I and III only (d) I, II and III

30. Although it is permitted to run the grounding conductor outside of the conduit, it is safer to run it with the power conductors to provide better protection in case of ____.

(a) mechanical damage **(b) a lightning strike**
(c) a ground fault **(d) a second ground fault**

31. The feeder for six 20 amp receptacles supplying shore power shall be calculated at ____ percent of the sum of the rating of the receptacles.

(a) 70 (b) 80 (c) 90 (d) 100

32. ____ conductors shall be used for wiring on fixture chains and other movable parts.

(a) Solid (b) Covered (c) Insulated (d) Stranded

33. The rating of a lampholder on a circuit which operates at a voltage less than 50 volts shall be at least ____ watts.

(a) 220 (b) 660 (c) 330 (d) 550

34. Health care low voltage equipment frequently in contact with bodies of persons shall not exceed ____ volts.

(a) 50 (b) 115 (c) 10 (d) 8

35. A manufactured assembly designed to support and energize lighting fixtures that are capable of being readily repositioned is ____.

(a) ceiling grid lighting **(b) electric discharge lighting**
(c) lighting track **(d) open circuit lighting**

36. Rigid metal conduit shall be permitted to be installed in concrete, in direct contact with the earth, or in areas subject to severe influences where protected by ____ and judged suitable for the condition.

(a) ceramic (b) corrosion protection (c) PVC (d) orangeburg

37. A park trailer is one built on a single chassis mounted on wheels and having a gross trailer area not exceeding ____ square feet in the set-up mode.

(a) 400 (b) 450 (c) 500 (d) 600

38. Nonmetallic extensions shall be supported every ____ inches with an allowable 12" to the first fastening where the connection to the supplying outlet is by means of an attachment plug.

(a) 6 (b) 8 (c) 10 (d) 16

39. The National Electrical Code is ____.

(a) intended to be a design manual
(b) meant to be used as an instruction manual for untrained persons
(c) the practical safeguarding of persons and property
(d) published by Bureau of Standards

40. Which of the following is/are correct about open wire systems on insulators?

I. surface-type snap switches do not need boxes
II. conductor supports shall be within 6" of a tap
III. surface-type snap switches shall be mounted on insulating material

(a) I only (b) II only (c) I and III only (d) I, II and III

41. Optical fiber cables transmit light for ____ through an optical fiber.

(a) communications (b) signaling (c) control (d) all of these

42. The equipment bonding jumper on the supply side of the service is sized by the rating of the ____.

(a) overcurrent protective device
(b) service entrance conductors
(c) service drop
(d) load to be served

43. Transformers over 112 1/2 kva shall not be within ____ of combustible material.

(a) 12" (b) 16" (c) 24" (d) 30"

44. How many #12 XHHW conductors can you install in a 3/8" flexible metal conduit with the use of inside fittings?

(a) 1 (b) 2 (c) 3 (d) 0

45. The overall covering of UF cable shall be ____.

I. suitable for direct burial in the earth
II. flame-retardant
III. moisture, fungus and corrosion resistant

(a) III and II only (b) I only (c) I and III only (d) I, II and III

46. In a Class II location, where electrically-conducting dust is present, flexible connections at motors could be made with ____.

(a) flexible metal conduit
(b) type AC armored cable
(c) hard usage cable
(d) liquid-tight flexible metal conduit with listed fittings

47. The short-circuit and ground-fault protective device protecting the branch circuit shall have sufficient ____ to permit the motor-compressor to start.

(a) voltage (b) current (c) time delay (d) capacity

48. Where magnesium, aluminum or aluminum-bronze powders may be present, transformers ____.

(a) must be dust-tight
(b) may be pipe-ventilated
(c) must be approved for Class II, Division I
(d) are not allowed

49. Knife switches rated for more than 1200 amperes at 250 volts ____.

(a) are used only as isolating switches
(b) may be opened under load
(c) should be placed so that gravity tends to close them
(d) should be connected in parallel

50. All electric equipment, including power supply cords, used with storable pools shall be protected by ____.

(a) fuses (b) circuit breakers (c) double-insulation (d) GFCI

51. What is the minimum working clearance on a circuit 277 volts to ground, exposed live parts on one side and grounded parts on the other side of the working space?

(a) 3 feet (b) 3 1/2 feet (c) 4 feet (d) 6 feet

52. Where the calculated number of conductors, all of the same size, includes a decimal fraction, the next higher whole number shall be used where this decimal is ____ or larger.

(a) .5 (b) .08 (c) .008 (d) .8

53. An oil switch is used on a motor circuit whose rating ____ or 100 amps.

(a) is 600 volts **(b) exceeds 600 volts**
(c) does not exceed 600 volts **(d) none of these**

54. The conductor between a lightning arrester and the line for installations operating at 1000 volts or more must be at least ____ copper.

(a) #14 (b) #6 (c) #8 (d) #12

55. Electric sign enclosures may be constructed of wood if ____.

(a) kept 1" from lampholders **(b) wood is not permitted at all**
(c) kept 2" from lampholders **(d) none of these**

56. Selecting a primary ____ location to achieve the shortest practical primary protector grounding conductor will help limit potential differences between communications circuits and other metallic systems.

(a) limiter (b) protector (c) neutral (d) common

57. Fuses, circuit breakers, or combinations thereof, shall not be connected in parallel except ____.

I. field installed II. factory-assembled in parallel, and listed as a unit III. readily accessible

(a) I only (b) II only (c) III only (d) I and III

58. Electric-discharge tubing operating at more than 7500 volts shall clear the adjacent surface by not less than ____.

(a) 1/4" (b) 1/2" (c) 1" (d) 3"

59. The ampacity of the phase conductors from the generator terminals to the first overcurrent device shall not be less than ____ percent of the nameplate rating of the generator.

(a) 75 (b) 115 (c) 125 (d) 140

60. Flexible cord shall not be used as a substitute for the ____ wiring of a structure.

(a) temporary (b) fixed (c) concealed (d) none of these

61. Electrodes of pipe or conduit shall not be smaller than ____.

(a) 1/2" (b) 5/8" (c) 3/4" (d) 1"

62. A night club lighting dimmer installed in an ungrounded conductor shall have overcurrent protection rated at no more than ____ percent.

(a) 50 (b) 70 (c) 80 (d) 125

63. A motel conference room is designed for the assembly of 100 or more persons. The room is fire-rated construction. One of the following wiring methods shall be required:

(a) rigid nonmetallic conduit
(b) MI cable
(c) nonmetallic sheathed cable
(d) type NMB cable

64.Busways rated over 600 volts shall have all conductor termination and connection hardware accessible for ____.

(a) installation (b) connection (c) maintenance (d) all of these

65. In areas where walls are frequently washed, conduit should be mounted with ____ inches of air space between the wall and the conduit.

(a) 1/8" (b) 1/4" (c) 3/8" (d) 1/2"

66. #12 aluminum conductors completely enclosed in the motor control enclosure shall be permitted to be protected with an overcurrent device where the rating is not more than ____ amperes.

(a) 20 (b) 80 (c) 100 (d) 125

67.Cable tray systems shall not be used in ____ or where subject to severe physical damage.

(a) tunnels (b) hoistways (c) hazardous locations (d) 600 volt systems

68. What is the minimum burial depth for rigid nonmetallic conduit in a dispensing station Class I, Division I location?

(a) 18" (b) 24" (c) 30" (d) cannot be used in Class I, Division I location

69. Conductors ____ and larger shall be stranded when installed in raceways.

(a) #12 (b) #10 (c) #8 (d) none of these

70. The minimum size service for a mobile home in a mobile home park is ____ amps.

(a) 80 (b) 70 (c) 200 (d) 100

2005

OPEN BOOK EXAM #16

70 QUESTIONS
TIME LIMIT - 2 HOURS

TIME SPENT ☐ **MINUTES**

SCORE ☐ **%**

MASTER OPEN BOOK EXAM #16 Two Hour Time Limit

1. In damp or wet locations, cabinets and cutout boxes of the surface type shall be mounted so there is at least ____ air space between the enclosure and the wall or other supporting surface.

(a) 6.35" (b) 6" (c) 3/4" (d) 1/4"

2. An 86 cubic inch junction box has two 1/2" rigid metal conduits entering the box. The conduits are supported within 36" of the box. The conduits enter the box through threaded hubs, one conduit through the top of the box. Select the correct statement concerning the Code requirements for this type of installation.

(a) The conduits must be supported within 18" of the box.
(b) The conduits must be supported within 12" of the box.
(c) The Code requires this box to be supported.
(d) The Code does not require this box to be supported.

3. The disconnecting means shall be ____ of the sign or outline lighting that it controls.

I. within sight of sign II. within 50' of sign III. within 75' of sign

(a) I only (b) II only (c) I and II only (d) I and III only

4. Thermoplastic insulation for fixture wires may be deformed at ____ temperatures where subjected to pressure, requiring care to be exercised during installation and at points of support.

I. 0 degrees F II. 72 degrees F III. 212 degrees F

(a) I only (b) II only (c) III only (d) II and III only

5. The installation of underfloor raceways shall be permitted ____.

I. beneath the surface of concrete
II. beneath floor material other than concrete
III. where laid flush with the concrete floor and covered with linoleum

(a) I only (b) II only (c) III only (d) I, II and III

6. FCC carpet squares that are adhered to the floor shall be attached with ____.

(a) tacking strip (b) release-type adhesive (c) glue (d) none of these

7. Electric equipment with a metal enclosure or with a nonmetallic enclosure listed for the use and having adequate ____ producing characteristics, and associated wiring material suitable for the ambient temperature shall be permitted to be installed in other space used for environmental air unless prohibited in this Code.

I. fire-resistant II. low-smoke

(a) I or II (b) I and II (c) I only (d) II only

8. Conductors shall be ____ to provide ready and safe access in underground and subsurface enclosures, into which persons enter for installation and maintenance.

(a) readily accessible (b) exposed (c) racked (d) enclosed

9. What is the ampacity of four #6 THW copper current-carrying conductors enclosed in schedule 80 PVC conduit, 8 feet in length entering a trench?

(a) 65 amps (b) 52 amps (c) 44 amps (d) 40 amps

10. Control circuit devices with screw-type pressure terminals used with a #14 or smaller copper conductors shall be torqued to a minimum of ____ pound-inches unless identified for a different torque value.

(a) 5 (b) 6 (c) 7 (d) 8

11. A metal pole supporting a light fixture shall have an accessible handhole not less than ____ inches having a raintight cover shall provide access to the supply terminations within the pole or pole base.

(a) 2 x 6 (b) 3 x 4 (c) 2 x 4 (d) 4 x 8

12. Liquidtight flexible metal conduit may be used in which of the following locations?

(a) in areas that are both exposed or concealed
(b) in areas where the ambient temperature is to be greater than 194 degrees C
(c) in areas that are subject to physical damage
(d) in connection areas for gasoline dispensing pumps

13. The minimum number of receptacles in a patient bed location of a hospital general care area should be ____.

(a) one (b) two (c) three (d) four

14. Surge arrester grounding conductors that run in metal enclosures should be ____.

(a) bonded on one end of the enclosure only **(b) bare**
(c) bonded at both ends of such enclosure **(d) insulated**

15. On a recreational vehicle, the chassis rail and any welded addition thereto of metal thickness of ____ or greater shall be used for the frame.

(a) .025 (b) .053 (c) .065 (d) .072

16. Heating panels installed under floor covering, shall not exceed ____ watts per square foot of heated area.

(a) 6 (b) 8 (c) 12 (d) 15

17. Where one side of the motor control circuit is grounded, the motor control circuit shall be so arranged that an ____ ground in the remote-control devices will not start the motor.

(a) intentional (b) accidental (c) isolated (d) low-voltage

18. The disconnecting means for a motor shall be a ____.

I. motor-circuit switch rated in horsepower II. circuit breaker III. branch circuit

(a) I only (b) II only (c) III only (d) I or II

19. Plate electrodes of nonferrous metal shall be at least ____ inches in thickness.

(a) 0.006 (b) 0.06 (c) 0.25 (d) 0.75

20. A discharge circuit of a capacitor shall ____.

I. be permanently connected to the terminals of the capacitor
II. be provided with automatic means of connecting it to the terminals of the capacitor bank after disconnection of the capacitor from the source of supply

(a) I only (b) II only (c) both I and II (d) either I or II

21. Emergency illumination shall include all required means of ____ and all other lights specified as necessary to provide required illumination.

I. illuminated exit signs II. egress lighting III. show window lighting

(a) I only (b) II only (c) I and II only (d) I, II and III

22. Which of the following may be used to insure continuity of the equipment grounding conductor in a branch circuit?

I. Connections to lampholders II. Terminals of receptacles

(a) I only (b) II only (c) both I and II (d) neither I nor II

23. Pendant fixtures shall be suspended in a Class I location by rigid non-metallic conduit if ____.

I. not more than 12" long II. braced if over 12" long

(a) I only (b) II only (c) both I and II (d) neither I nor II

24. Individual overcurrent protection for an appliance panelboard may be omitted if the panelboard feeder has overcurrent protection not greater than ____.

(a) the combined capacity of the overcurrent device in the panelboard
(b) 200 amps
(c) the rating of the panelboard
(d) the feeder ampacity

25. Power and control tray cable can be installed ____.

I. as open cables on brackets or cleats II. in cable trays in hazardous locations

(a) I only (b) II only (c) both I and II (d) neither I nor II

26. For a legally required standby system, power will be available for the application within ____.

(a) 30 seconds (b) 10 seconds (c) 2 minutes (d) one minute

27. ___ of the following is a true statement.

I. Wiring above a nonfire rated ceiling shall not be supported by the ceiling support wires.
II. Wiring above a fire rated ceiling shall not be supported by the ceiling support wires.
III. Cables and raceways shall not be permitted to be supported by ceiling grids.
IV. The ceiling support wires shall be permitted to support branch circuit wiring if in accordance with the ceiling manufacturer's instructions.

(a) I & II only (b) II & III only (c) IV only (d) I, II, III & IV

28. 4160v feeder, in no case shall the fuse rating in continuous amperes exceed three times, or the long-time trip element setting of a breaker ____ times, the ampacity of the conductor.

(a) 3 (b) 4 (c) 6 (d) 8

29. Conductors installed in rigid nonmetallic conduit operating at 56 kv, covered with 2" of concrete shall have a minimum cover of ____ inches.

(a) 30 (b) 18 (c) 24 (d) 42

30. The disconnecting means for portable X-ray equipment operating on 120v, 30 ampere or less branch circuits may be a ____.

I. general duty snap switch loaded not more than 80%
II. grounding type attachment plug and receptacle

(a) I only (b) II only (c) both I and II (d) neither I nor II

31. A transformer case need not be grounded if ____.

I. inside a fenced enclosure II. mounted 12' up a wood pole

(a) I only (b) II only (c) both I and II (d) neither I nor II

32. An auxiliary gutter shall not extend a greater distance than ____ feet beyond the equipment which it supplements.

(a) 10 (b) 20 (c) 30 (d) 40

33. The power supply lines for portable switchboards in a theater shall be sized considering the neutral as a ____.

(a) non-current carrying conductor **(b) current-carrying conductor**
(c) bonding conductor **(d) none of these**

34. Bonding together of all separate grounding electrodes will limit ____ between them and between their associated wiring systems.

(a) potential differences (b) high frequencies (c) stray currents (d) arcing

35. The fuse is designed so that discharged gases will not ignite or damage insulation in the path of the discharge or propagate a flashover to or between grounded members or conduction members in the path of the discharge when the distance between the vent and such insulation or conduction members conforms to manufacturer's recommendations. This is called a ____.

(a) nonvented power fuse **(b) controlled vented power fuse**
(c) expulsion fuse **(d) power fuse unit**

36. The voltage limitation for electrical nonmetallic tubing is ____ volts.

(a) 600 (b) 500 (c) 450 (d) 300

37. Mobile home disconnecting means shall be located not less than ____ feet above finished grade or working platform.

(a) 8 (b) 6 (c) 4 (d) 2

38. Listed baseboard heaters include instructions which may not permit their installation ____ dwelling unit receptacle outlets.

(a) above (b) below (c) beside (d) adjacent

39. In commercial garages all 125v single-phase, 15 and 20 amp receptacles, where ____ are to be used, shall provide GFCI protection for personnel.

I. portable lighting devices
II. electrical hand tools
III. electrical automotive diagnostic equipment

(a) I only (b) II only (c) III only (d) I, II and III

40. Transformers and transformer vaults shall be ____ to qualified personnel for inspection and maintenance.

(a) accessible (b) readily accessible (c) externally operable (d) none of these

41. Joints in festoon wiring shall be ____.

(a) spliced
(b) staggered
(c) double-insulated
(d) soldered

42. The allowable distance between supports of nonmetallic-sheathed cable installed in an on site constructed, one-family dwelling is a maximum ____ feet.

(a) 2 (b) 3 1/2 (c) 4 (d) 4 1/2

43. The ____ shall not be less than the noncontinuous load plus 125% of the continuous load.

(a) branch-circuit rating (b) continuous load (c) non-continuous load (d) conductor size

44. Double-pole switched lampholders supplied by the ungrounded conductors of a circuit, the switching device of lampholders of the switched type shall ____ disconnect both conductors of the circuit.

(a) separately (b) simultaneously (c) readily (d) automatically

45. Where there is no patient equipment grounding point, it is important that the distance between the reference grounding point and the patient vicinity be ____ to minimize any potential differences.

(a) 6 feet (b) 8 feet (c) 10 feet (d) as short as possible

46. Exposed non-current carrying metal parts of fixed equipment must be grounded under all of the following conditions except ____.

(a) where supplied by metal clad wiring
(b) electrically heated devices with the frames insulated from ground
(c) where in electrical contact with metal
(d) where located in damp locations

47. The general rule is half-round and flat-top raceways not over ____ in width shall have not less than 3/4" of concrete or wood above the raceway.

(a) 2" (b) 3" (c) 4" (d) 6"

48. A 120v lighting circuit and a 277v lighting circuit are installed in the same raceway, the 120v circuit has a white colored grounded conductor, the 277v grounded conductor would be ____.

(a) white (b) gray (c) white with a green stripe (d) white with a yellow stripe

49. Flexible cords for aircraft energizers and ground support equipment shall ____.

I. be approved for extra-hard usage
II. include a grounded conductor
III. be approved for the type of service

(a) I only (b) I and III only (c) I and II only (d) I, II and III

50. In Class I, Zone 0 locations, only intrinsically safe wiring methods in accordance with ____ shall be permitted.

(a) UL (b) NEMA (c) Article 504 (d) Article 725

51. A metallic water system is used as a grounding means. Which of the following statements is/are true?

I. must be bonded around meter II. must bond around insulated joints

(a) I only (b) II only (c) both I and II (d) neither I nor II

52. Conductors installed outdoors at 7200v, would have a minimum clearance phase to phase of ____.

(a) 4 1/2" (b) 4" (c) 7" (d) 5 1/2"

53. The inside diameter area of square inch for a 1 5/8" x 3 1/4" strut-type channel raceway is ____.

(a) 3.169 (b) 4.308 (c) 4.803 (d) 2.780

54. Power-limited fire protective signaling circuits where installed in exposed cable, located within 7 feet of the floor, cable shall be securely fastened in an approved manner at intervals of not more than ____ inches.

(a) 18 (b) 24 (c) 30 (d) 36

55. Conductors operating at less than 50 volts and not covered in Articles 650, 725 and 760, shall not be smaller than ____ copper or equivalent.

(a) #16 (b) #18 (c) #14 (d) #12

56. Armored cable installed in thermal insulation shall have conductors rated at ____. The ampacity of the cable installed in these applications shall be that of 60 degree C conductors.

(a) 60 degrees C (b) 194 degrees F (c) 75 degrees C (d) 90 degrees F

57. Indoor antennas and indoor lead-ins shall be permitted to occupy the same box or enclosure with conductors of other wiring systems where separated from such other conductors by an effective permanently installed ____.

(a) wall (b) divider (c) insulator (d) barrier

58. Components of lighting track systems of different voltages shall not be ____.

(a) connected (b) interchangeable (c) polarized (d) none of these

59. Where separate services supply a building and are required to be connected to a grounding electrode, ____ shall be used.

(a) multiple electrodes **(b) the same grounding electrode**
(c) two electrodes **(d) paralleled electrodes**

60. Snap switches can be grouped or ganged in outlet boxes if voltages between adjacent switches do not exceed ____ volts.

(a) 150 (b) 300 (c) 350 (d) 400

61. Excluding fuses and exceptions, how many overcurrent protection devices, such as trip coils, relays or thermal cutouts, are required on a three-phase motor?

(a) 0 (b) 1 (c) 2 (d) 3

62. All of the following sizes of solid aluminum conductors can be made of aluminum alloy except ____.

(a) #14 AWG (b) #12 AWG (c) #10 AWG (d) #8 AWG

63. Lighting fixtures installed over vehicle lanes inside a commercial garage shall be installed a minimum of ____ feet.

(a) 8 (b) 10 (c) 12 (d) 15

64. The bare neutral of aluminum conductors may not be used underground on a service except ____.

(a) if protected at not more than 20a
(b) if protected with oxide inhibiter
(c) if installed in aluminum conduit
(d) where part of a cable assembly identified for direct burial

65. Which of the following statements about circuit breakers used to provide load protection for a high voltage motor circuit is correct?

I. The breakers shall simultaneously disconnect all ungrounded conductors to the motor.
II. The breakers may sense a fault current by means of integral external sensing elements.

(a) I only (b) II only (c) both I and II (d) neither I nor II

66. The maximum allowable rating for overcurrent devices for an AC transformer welder is ____ percent of the rated primary current.

(a) 100 (b) 125 (c) 150 (d) 200

67. In a Class I, Division II area, bonding can be accomplished by ____.

I. double locknuts   II. locknut-bushings

(a) I only   (b) II only   (c) both I and II   (d) neither I nor II

68. In a grounded system the grounding electrode conductor shall be connected to the following ____.

I. at any accessible point from the load end of the service drop
II. including the terminal to which the grounded service conductor is connected at the service disconnecting means

(a) I only   (b) II only   (c) both I and II   (d) neither I nor II

69. Two conductors in a rigid PVC schedule 40 conduit, what is the percent of allowable fill?

(a) 31%   (b) 40%   (c) 53%   (d) 80%

70. An accessible means external to enclosures for connecting intersystem bonding and grounding conductors shall be provided at the service by at least one of the following means ____.

I. exposed grounding electrode conductor
II. exposed nonflexible metallic service raceways
III. approved means for the external connection of a copper or other corrosion-resistant bonding, or grounding conductor to the service raceway or equipment

(a) I only   (b) II only   (c) III only   (d) I, II or III

ANSWERS

ANSWERS MASTER CLOSED BOOK EXAM #1

1. **(d)** separately-excited
2. **(b)** impedance
3. **(d)** 120°
4. **(c)** efficiency
5. **(a)** the number of lines of flux
6. **(b)** impregnated paper
7. **(a)** greater
8. **(a)** resistance
9. **(d)** temporary magnet
10. **(c)** carbon
11. **(d)** friction
12. **(b)** E2/W
13. **(b)** E x I x T
14. **(b)** decreases
15. **(a)** double
16. **(c)** 1/1000
17. **(d)** 6530cm
18. **(c)** element
19. **(c)** valence
20. **(c)** 9Ω
21. **(a)** 10 times
22. **(a)** utilize energy
23. **(d)** switch
24. **(d)** output divided by input
25. **(c)** mhos
26. **(a)** transformer
27. **(d)** 50 feet of wire
28. **(d)** result in damage to ballast
29. **(a)** appliances independent
30. **(d)** 2Ω
31. **(a)** excess of electrons
32. **(d)** charges
33. **(b)** different current ratings
34. **(c)** change in the voltage
35. **(c)** hardened steel
36. **(b)** voltage applied
37. **(c)** infinite
38. **(b)** induction
39. **(b)** 40°C
40. **(b)** refraction
41. **(c)** inductive load
42. **(b)** prevent chemical reactions
43. **(a)** delays the change in current
44. **(c)** AC easily changed
45. **(c)** peak
46. **(a)** reactive power is decreased
47. **(c)** less than that of
48. **(a)** they lubricate and polish
49. **(b)** high starting torque
50. **(b)** the same as

ANSWERS MASTER CLOSED BOOK EXAM #2

1. **(c)** increase in vd
2. **(a)** tool to bend pipe
3. **(c)** oil a **sleeve** bearing
4. **(a)** loose connection
5. **(a)** six 230.71(A)
6. **(d)** vacuum or gas
7. **(a)** cabinet DEF 100
8. **(c)** voltmeter III only
9. **(d)** greater at partial loads
10. **(b)** intermittent DEF 100
11. **(b)** receptacle outlet DEF 100
12. **(c)** covered DEF 100
13. **(c)** covered by Code 110.13(B)
14. **(b)** authority DEF 100
15. **(b)** emergency system 700.1
16. **(b)** Chapter 5
17. **(a)** yes 300.3(C1)
18. **(a)** load
19. **(c)** RHW Table 310.13
20. **(c)** poor neutral connection
21. **(a)** reduce 240.2 DEF
22. **(d)** service DEF 100
23. **(b)** DEF 100
24. **(a)** continuous **load** DEF 100
25. **(d)** the contact resistance
26. **(b)** piezoelectricity
27. **(b)** higher voltage
28. **(c)** rectifier
29. **(a)** delta
30. **(c)** III & IV only
31. **(d)** 1000 $W = I^2R$
32. **(b)** will not
33. **(d)** high
34. **(c)** compensate for vd
35. **(c)** largest
36. **(b)** main bonding jumper DEF 100
37. **(a)** good connection 110.14(A)
38. **(d)** megger
39. **(a)** protect user
40. **(a)** control speed of motor
41. **(d)** ammeter, wattmeter & voltmeter
42. **(a)** 25 or 60 cycle
43. **(b)** blades dead
44. **(b)** lead sheath
45. **(d)** damaging the finish
46. **(b)** rubber
47. **(d)** low wattage-high resistance
48. **(b)** commutator or slip rings
49. **(c)** 24 volts
50. **(d)** 720 watts

ANSWERS MASTER CLOSED BOOK EXAM #3

1. **(c)** counterpoise
2. **(d)** impulse
3. **(c)** flashpoint
4. **(d)** line side of service
5. **(d)** any of these
6. **(d)** insulation resistance test
7. **(c)** be decreased
8. **(d)** oscilloscope
9. **(b)** zero
10. **(b)** load
11. **(a)** 0.50
12. **(d)** insulation is not the only protection
13. **(a)** 40.66 volts
14. **(a)** zero
15. **(b)** conductor
16. **(a)** keeper
17. **(c)** two
18. **(b)** rigid non-metallic conduit
19. **(b)** interlock
20. **(b)** bimetallic
21. **(c)** low voltage
22. **(c)** utilize energy
23. **(d)** four times
24. **(b)** semiconductors
25. **(d)** current proportional
26. **(b)** magnetomotive
27. **(c)** low side
28. **(a)** zero
29. **(d)** wiring
30. **(c)** inductance
31. **(a)** elastance
32. **(c)** toggle switch
33. **(a)** phase volts equal line volts
34. **(c)** current capacity of cells
35. **(d)** inching
36. **(c)** reciprocal of frequency
37. **(d)** both inductively and conductively
38. **(b)** be decreased
39. **(d)** maximum value
40. **(b)** continuity
41. **(b)** rotor
42. **(c)** thermocouple
43. **(d)** I^2R losses are lower
44. **(a)** alternating current
45. **(a)** sustained overload
46. **(c)** breakdown
47. **(a)** cleat
48. **(c)** high inertia
49. **(d)** 35 volts
50. **(d)** electrostatic force

ANSWERS MASTER CLOSED BOOK EXAM #4

1. **(c)** greater than the phase voltage
2. **(d)** Article 630 welders
3. **(b)** Labeled DEF 100
4. **(a)** to reduce inductive heat
5. **(a)** equal to the secondary phase current
6. **(a)** two
7. **(d)** medium base
8. **(b)** pole of a switch
9. **(a)** 10' 230.26
10. **(a)** reversed 110.57
11. **(d)** orange triangle 406.2(D)
12. **(b)** armature current is constant
13. **(c)** I and II only
14. **(a)** enclosed DEF 100
15. **(a)** isolated DEF 100
16. **(c)** Class III 500.5(D)
17. **(c)** Chapter 8 90.3
18. **(d)** 90% efficiency
19. **(d)** 6 steps 210.70(A2c)
20. **(d)** Article 760 fire alarm
21. **(b)** negatively
22. **(b)** wye connected
23. **(d)** Hysteresis losses
24. **(c)** neutral carries unbalance
25. **(b)** intermittent duty
26. **(a)** #6 or smaller 200.6(A)
27. **(c)** either I or II 314.25(A)
28. **(b)** I & II only 410.4
29. **(d)** Article 800
30. **(b)** Article 490
31. **(b)** I & III only 410.65(C) & 410.73(E1)
32. **(d)** I,II & III 300.3(B)
33. **(c)** either I or II
34. **(c)** watertight DEF 100
35. **(b)** 3 amps 200/5= 4 ratio 120/4 = 3a
36. **(c)** burnished
37. **(c)** score the conduit during the pull
38. **(b)** power factor meter
39. **(a)** AC current flows
40. **(a)** fixture wire 410.16(C)
41. **(b)** 31% csa Chaper 9 Table 1
42. **(a)** 6 feet 210.52
43. **(a)** metal water pipe 250.53(D2)
44. **(b)** 1/8" 314.21
45. **(d)** I,II, & III 110.3(B) 430.7(C)
46. **(a)** one circuit
47. **(a)** high resistance
48. **(d)** 8Ω current flows same in series
49. **(a)** 2Ω current divides in parallel
50. **(a)** mandrel

ANSWERS MASTER CLOSED BOOK EXAM #5

1. **(b)** loss depends on va
2. **(b)** voltage across the same
3. **(b)** current required is higher
4. **(c)** transformer
5. **(c)** interlock was welded
6. **(b)** intentionally grounded
7. **(a)** remains the same
8. **(a)** one valence electron
9. **(b)** power is zero
10. **(b)** separated by dielectric
11. **(d)** interrupting
12. **(d)** normally open
13. **(d)** 1/12 cycle
14. **(a)** the air gap
15. **(c)** segments are short-circuited
16. **(b)** flashover
17. **(a)** aids the rotor flux
18. **(b)** bonding
19. **(d)** all of these
20. **(d)** controller
21. **(c)** power panelboard
22. **(d)** breather effectiveness
23. **(b)** trip-free
24. **(a)** polarity
25. **(d)** 1840
26. **(c)** capacitor in parallel with contacts
27. **(d)** compensator
28. **(d)** conductance
29. **(d)** electronic
30. **(d)** four times as great
31. **(c)** likely to burn out quickly
32. **(d)** varying magnetic field
33. **(c)** Joules
34. **(b)** megger
35. **(c)** higher fault current, shorter time
36. **(a)** has better speed control
37. **(c)** pull box
38. **(a)** infinite resistance
39. **(a)** extend range, reduce resistance
40. **(d)** air gap
41. **(a)** coulomb
42. **(b)** one ampere in both resistors
43. **(c)** retentivity
44. **(c)** air terminal
45. **(d)** insulation resistance
46. **(d)** infinite resistance and zero inductance
47. **(c)** synchronous
48. **(c)** draw excessive current and overheat
49. **(d)** peak load
50. **(d)** bonding

ANSWERS MASTER CLOSED BOOK EXAM #6

1. **(a)** transformer is associated with AC
2. **(c)** silver for relay contacts
3. **(d)** controller DEF 100
4. **(a)** copper-clad 10% DEF 100
5. **(b)** an overload DEF 100
6. **(b)** high power factor close to unity
7. **(b)** transformer core is steel
8. **(a)** adjustable DEF 100
9. **(c)** signals 430.71
10. **(a)** voltage is emf
11. **(d)** Δ delta symbol
12. **(c)** 4 1/2' 348.30(A)
13. **(b)** 4 ohms $R = E^2/W$
14. **(c)** good conductors free electrons
15. **(b)** II and IV only 200.7C1
16. **(c)** 0.0129Ω half as much resistance
17. **(a)** terminal connected to grounded wire
18. **(b)** B phase
19. **(d)** III and IV only 250.4(A2) & (B4)
20. **(d)** Article 430
21. **(d)** ampacity DEF 100
22. **(a)** I,II and III chain wrench
23. **(c)** nipple may or may not be threaded
24. **(d)** covered conductor DEF 100
25. **(b)** fitting DEF 100
26. **(d)** vertical position 240.33
27. **(a)** ungrounded conductor O.C.P. 240.20(A)
28. **(b)** more than 6' 210.52(A1)
29. **(b)** intentional ground DEF 100
30. **(c)** octagonal box for ceiling outlet
31. **(b)** soldered fittings 250.70
32. **(b)** I and II only 665.1
33. **(c)** VD 4.17% 5v/120v
34. **(d)** insulation
35. **(b)** II only service overhead DEF 100
36. **(b)** current
37. **(a)** I,II and III wire, solder and lug
38. **(c)** RHH 90° rating Table 310.13
39. **(c)** 20 amps series-parallel circuit
40. **(c)** score the conduit during the pull
41. **(d)** I, II and III increases resistance
42. **(c)** 120 mechanical degrees
43. **(c)** rigid PVC 18" cover Table 300.5
44. **(d)** nipple 60% fill Chap. 9 Table 1 note 4
45. **(a)** 1 conductor 314.16(B5) equip. grd.
46. **(c)** symbol for thermal overload
47. **(a)** one residence hallway 210.52(H)
48. **(d)** I, II or III 220.14(I) 180va per yoke
49. **(b)** Article 770
50. **(c)** pressure connector DEF 100

ANSWERS MASTER CLOSED BOOK EXAM #7

1. **(d)** does not affect the performance
2. **(a)** less reactance
3. **(c)** has fewer electrons
4. **(b)** converts an electrical input into action
5. **(c)** degaussing
6. **(c)** AC high potential
7. **(d)** extract moisture in the air
8. **(b)** programmable logic controller
9. **(a)** Joule
10. **(a)** vault
11. **(a)** star-delta
12. **(a)** branch circuit conductors
13. **(c)** flashover
14. **(b)** span
15. **(d)** 1600
16. **(b)** increase, reduce
17. **(b)** zero
18. **(b)** lagging
19. **(a)** voltage
20. **(d)** mercury lamp
21. **(a)** reactor
22. **(a)** no voltage release
23. **(d)** 1373
24. **(a)** resistance
25. **(a)** hysteresis loss
26. **(a)** decreases
27. **(d)** maximum
28. **(b)** 100w, 240v
29. **(b)** an inductor
30. **(b)** pilot
31. **(d)** incandescent
32. **(c)** period
33. **(a)** diameter of the wire
34. **(a)** cosine of the phase angle
35. **(b)** inverter
36. **(c)** reverses the magnetic polarity
37. **(c)** a sparkover
38. **(d)** receptacle
39. **(c)** both commutator and slip ring
40. **(a)** at the start of current flow
41. **(a)** reactive power
42. **(b)** 500Ω, 2w
43. **(c)** voltage
44. **(c)** cathodic
45. **(b)** auxiliary windings
46. **(b)** resistance
47. **(b)** ampere-turns and voltage per turn
48. **(c)** skin effect
49. **(a)** talc
50. **(b)** additive

ANSWERS MASTER CLOSED BOOK EXAM #8

1. **(a)** expansion joints
2. **(b)** interrupting
3. **(d)** multiply strands
4. **(d)** 5' 680.6(2)
5. **(b)** commutator bar separators
6. **(d)** not to protect the end of wire
7. **(b)** 864 watts
8. **(d)** safe for stepping down
9. **(d)** XC
10. **(b)** electrical energy only
11. **(c)** insulation resistance
12. **(d)** controller DEF 100
13. **(d)** longer bulb life
14. **(c)** apparent
15. **(b)** 480v
16. **(c)** all in parallel
17. **(b)** voltage difference
18. **(b)** current the same in series
19. **(b)** moisture proofing
20. **(b)** prevent corrosion
21. **(b)** I and II only 660.1
22. **(c)** low resist. & low melting point
23. **(d)** all DEF 100
24. **(a)** resistor
25. **(c)** asbestos
26. **(a)** fuse clips are too loose
27. **(c)** fitting
28. **(b)** inductance
29. **(c)** effective difference
30. **(c)** turns-ratio
31. **(d)** 5000 amps or more
32. **(a)** magnetic effect
33. **(d)** 1,000,000
34. **(c)** loose connection
35. **(d)** switch
36. **(a)** not permanently enclosed DEF 100
37. **(c)** farad
38. **(c)** even spacing, numerous lights
39. **(c)** one hot leg is shut off
40. **(a)** one
41. **(d)** Lightning arresters
42. **(c)** transformer
43. **(b)** oscilloscope
44. **(b)** aluminum
45. **(a)** 90°
46. **(a)** XL
47. **(a)** insulation not the only protection
48. **(d)** polarity
49. **(b)** R/Z
50. **(a)** 500 watts

ANSWERS MASTER CLOSED BOOK EXAM #9

1. **(b)** interrupting
2. **(b)** low
3. **(d)** all of these
4. **(a)** very high
5. **(d)** natural air cooled
6. **(a)** high voltage winding
7. **(d)** effective value
8. **(a)** low
9. **(c)** half of one wire
10. **(c)** DC resistance
11. **(d)** corona
12. **(a)** pilot
13. **(c)** input to electrical output
14. **(b)** pure unmagnetized iron
15. **(b)** dangerous under certain conditions
16. **(a)** speed, frequency, number of poles
17. **(c)** 170 volts
18. **(c)** kvar
19. **(d)** parrallel, series
20. **(d)** instantaneous-trip
21. **(a)** approximately no resistance
22. **(a)** metal halide
23. **(a)** series
24. **(b)** AC only
25. **(c)** direct on line
26. **(b)** thermocouple
27. **(d)** all of these
28. **(a)** 43.81 amps
29. **(c)** power
30. **(d)** reluctance
31. **(a)** capacitor
32. **(a)** high internal resistance
33. **(d)** resistive
34. **(c)** number of phases
35. **(b)** megger
36. **(b)** stator
37. **(a)** reverses in direction
38. **(a)** melting alloy
39. **(a)** one
40. **(d)** effective
41. **(a)** plugging
42. **(b)** protect from power reversal
43. **(b)** battery
44. **(a)** reduce voltage at starting
45. **(b)** 75%
46. **(c)** wire wound
47. **(a)** high
48. **(a)** ground terminal
49. **(d)** separately derived
50. **(d)** tie

ANSWERS MASTER CLOSED BOOK EXAM #10

1. **(c)** electro
2. **(a)** switches 1 and 3
3. **(c)** tachometer
4. **(b)** b.c. protection
5. **(b)** askarel DEF 100
6. **(d)** % of applied voltage
7. **(c)** demand factor DEF 100
8. **(b)** pressure
9. **(d)** either I or II
10. **(c)** less than any one
11. **(c)** wires do not support the light
12. **(d)** I, II & III
13. **(d)** I, II or III
14. **(c)** simultaneously
15. **(c)** area surrounding
16. **(d)** both a & b
17. **(b)** pf meter
18. **(c)** covered DEF 100
19. **(d)** I,II,III & IV
20. **(d)** all DEF 100
21. **(d)** Varying
22. **(b)** low transformation
23. **(a)** Y
24. **(d)** thermal cutout
25. **(b)** short circuited
26. **(c)** ohms
27. **(c)** volt
28. **(a)** current flow develops heat
29. **(c)** 90.2
30. **(b)** soft iron
31. **(c)** shorted
32. **(b)** 10 ohm resistor
33. **(c)** conductivity
34. **(b)** thermocouple
35. **(d)** induction exceeds capacitance
36. **(b)** zero
37. **(b)** electron flow
38. **(b)** three-phase
39. **(a)** threads per inch
40. **(c)** shunt
41. **(a)** magnetism
42. **(c)** voltage
43. **(d)** 5/8" seal 501.5(C3)
44. **(c)** effective
45. **(b)** rheostat
46. **(a)** direct
47. **(c)** 3/4" 344.28
48. **(d)** commutator
49. **(c)** zero
50. **(c)** good conductors

ANSWERS MASTER CLOSED BOOK EXAM #11

1. **(d)** capacitance exceeds inductance
2. **(b)** thread is tapered
3. **(d)** either I or III 230.22
4. **(a)** commutator
5. **(b)** cable terminal
6. **(d)** 70.7v 100v x .707 = 70.7v
7. **(d)** 1680 watts
8. **(a)** one strand of the conductor
9. **(d)** highest when bulb is on
10. **(a)** remote-control
11. **(c)** 1/2 ohm
12. **(d)** I and II
13. **(b)** shorted
14. **(a)** leather gloves over rubber
15. **(a)** cartridge fuses
16. **(c)** same ampacity Table 310.16
17. **(a)** accidentally energized
18. **(d)** I, II & III
19. **(d)** electrolyte
20. **(c)** DC
21. **(c)** connect two sections
22. **(d)** rawl plugs
23. **(c)** B - C - A
24. **(d)** lack of isolation
25. **(c)** I, II & V
26. **(a)** device DEF 100
27. **(c)** reduce pitting of contacts
28. **(b)** quench the arc
29. **(a)** inversely proportional to csa
30. **(c)** frequency meter
31. **(a)** chemical
32. **(b)** henrys
33. **(a)** DC motor
34. **(a)** some sort of ballast
35. **(b)** less
36. **(a)** 40° C
37. **(c)** 120°
38. **(a)** conductor moving inside a magnetic
39. **(a)** approved sealing compound
40. **(d)** 1/240 second
41. **(c)** watt-hour
42. **(a)** mica
43. **(c)** I or II only 500.1
44. **(c)** eddy current loss at minimum
45. **(a)** downward
46. **(a)** impedance
47. **(a)** squirrel cage
48. **(a)** $R = V^2/W$
49. **(d)** 1000 amp turns
50. **(c)** 120 volts

ANSWERS MASTER CLOSED BOOK EXAM #12

1. **(d)** 120°
2. **(b)** passing a point per second
3. **(d)** hydrometer
4. **(c)** voltmeter
5. **(d)** expansion bolts
6. **(b)** less voltage drop
7. **(a)** .375
8. **(c)** average
9. **(b)** voltmeter, ohmmeter, ammeter
10. **(a)** burn at full brightness
11. **(a)** I, II and III 640.1
12. **(d)** grounding cond. DEF 100
13. **(a)** iron
14. **(a)** ruin the tip
15. **(a)** hanging fixture
16. **(a)** Weatherproof DEF 100
17. **(c)** reverse any two
18. **(d)** cannot be used in e.p.
19. **(c)** three
20. **(c)** capacitor
21. **(d)** 180va 220.14(L)
22. **(d)** strip sandpaper
23. **(c)** induction
24. **(b)** current
25. **(a)** Kirchhoff's law
26. **(c)** E = I x R
27. **(d)** clamped perpendicular
28. **(a)** prevent loosening under vibration
29. **(a)** primary **current transformer**
30. **(d)** heat sensing element
31. **(b)** ammeter
32. **(b)** in a DC circuit
33. **(d)** conductor
34. **(a)** shock or injury
35. **(a)** oil
36. **(d)** conductance
37. **(b)** concealed
38. **(b)** autotransformer
39. **(a)** may transmit shock
40. **(a)** metal undergrd. water pipe 250.53(D2)
41. **(c)** the feeder
42. **(a)** general purpose DEF 100
43. **(c)** ease of voltage variation
44. **(a)** parallel
45. **(a)** 0.002v
46. **(c)** Class III 500.5(D)
47. **(d)** Stranded 410.28(E)
48. **(a)** use a template
49. **(c)** of inductance
50. **(d)** 105 volts

ANSWERS MASTER CLOSED BOOK EXAM #13

1. **(a)** mandrel
2. **(a)** frequency remains constant
3. **(a)** cut off circuit from source
4. **(d)** persons familiar with construction,etc.
5. **(b)** grounded conductor DEF 100
6. **(a)** tin and lead
7. **(c)** 49.49 amperes 70a x .707
8. **(c)** 360 watts W = E x I
9. **(c)** raintight and arranged to drain 225.22
10. **(b)** 3ø 208v, 1ø 120v and 1ø 208v
11. **(d)** CO_2 fire extinguisher
12. **(b)** oil
13. **(c)** antenna systems 820.1
14. **(c)** 6" bus bar 366.23(A)
15. **(d)** prevent shock
16. **(a)** potential transformer and voltmeter
17. **(c)** 144 watts
18. **(c)** a neon tester
19. **(d)** 0.625 decimal equivalent for 5/8"
20. **(c)** two sources for emergency lighting
21. **(b)** Chapter 6 elevators
22. **(d)** 40% fill easier to pull wires
23. **(a)** pull tape to half its original width
24. **(a)** nichrome wire heater coils
25. **(d)** one fourth as much heat at 120v
26. **(b)** 40 volts
27. **(a)** symbol for wiring connected
28. **(b)** one-eight bend is 45°
29. **(a)** 4Ω consume the most power parallel
30. **(a)** exerting a heavy pull balance on ladder
31. **(b)** relay is an actuating control
32. **(c)** 4800 watts
33. **(d)** cycle counter for relays
34. **(c)** most wattage all in parallel
35. **(b)** carbon is poorest conductor
36. **(b)** receptacle counts as two in box fill
37. **(c)** polarized plug
38. **(a)** electrolyte sulphuric acid in water
39. **(a)** low voltage is 600 volts and under
40. **(a)** orangeburg pipe is nonmetallic
41. **(c)** 19,200 watts
42. **(a)** symbol for temperature switch
43. **(c)** least effect on VD is 50 or 60 hz
44. **(b)** high power factor
45. **(b)** 1 1/4" is the largest hand bender
46. **(d)** short circuit has a very low resistance
47. **(d)** fuse
48. **(d)** 400 watts
49. **(c)** power factor kw to kva relationship
50. **(b)** 90 volts with switch 4 closed

ANSWERS MASTER CLOSED BOOK EXAM #14

1. **(d)** NFPA
2. **(d)** plan
3. **(d)** reached quickly DEF 100
4. **(d)** mech function DEF 100
5. **(d)** unbalanced 310.15(B4)
6. **(d)** I, II or III DEF 100
7. **(b)** ammeter II only
8. **(b)** free of shorts 110.7
9. **(a)** Type A ballast
10. **(b)** 1/120
11. **(a)** real power
12. **(b)** interpoles
13. **(c)** authority 90.4
14. **(d)** continuously DEF 100
15. **(a)** remote-control DEF 100
16. **(c)** grounded T.110.26(A) condition 2
17. **(b)** three-phase 4-wire
18. **(d)** broken
19. **(c)** one-thousand circular mils
20. **(d)** stationary 550.2 DEF
21. **(c)** an outlet DEF 100
22. **(a)** Thermal protector DEF 100
23. **(a)** instructions 110.3B
24. **(a)** free from hazard 90.1(B)
25. **(d)** output of motor
26. **(d)** I, II, & III
27. **(a)** one wattmeter
28. **(d)** I, II & III
29. **(c)** Chapter 7
30. **(d)** temperature
31. **(c)** continuity
32. **(a)** in series with the load
33. **(c)** at least two wires
34. **(d)** wire binding screws
35. **(c)** Fine print notes
36. **(c)** both I & II DEF 100
37. **(d)** 90% efficient
38. **(d)** continuous **DUTY** DEF 100
39. **(c)** dust will not interfere DEF 100
40. **(d)** all of these
41. **(a)** guarded DEF 100
42. **(c)** service drop DEF 100
43. **(a)** dead front DEF 100
44. **(c)** is a constant value
45. **(d)** **ampacity** remains the same
46. **(d)** I, II or III
47. **(b)** #18 402.6
48. **(a)** watt output decreases
49. **(b)**
50. **(a)** 144 volts

ANSWERS MASTER OPEN BOOK EXAM #1

1.	**(c)**	30"	110.26(A2)
2.	**(b)**	8' long aluminum	250.52(A & B)
3.	**(b)**	#4 copper	430.96 T.250.122
4.	**(d)**	30 amps	600.5(B2)
5.	**(b)**	single phase	T.430.47
6.	**(b)**	protected display	410.101(C8)
7.	**(c)**	#10 cu-clad alum.	230.31(B) ex.
8.	**(a)**	#6 copper	T.250.122
9.	**(d)**	completely through	501.15(A4) ex.1
10.	**(a)**	#4 copper	T.250.66
11.	**(c)**	600v	362.12(6)
12.	**(a)**	adjacent to	550.32(A)
13.	**(a)**	12.5%	250.102(C)
14.	**(a)**	0.5958	Table 5
15.	**(b)**	80'	T.300.19(A)
16.	**(c)**	18"	250.64(A)
17.	**(a)**	Class I Div. 1	500.5(B)
18.	**(c)**	18'	225.18(4)
19.	**(a)**	Class I	505.6(C) & FPN
20.	**(d)**	None	210.52(B)
21.	**(a)**	6"	T.300.5
22.	**(b)**	300%	630.32(A)
23.	**(c)**	90°C	T.310.13 note 4
24.	**(c)**	90°C	T.310.16
25.	**(d)**	0%	T.310.15(B2) ex.3
26.	**(c)**	III only	630.31(B) FPN
27.	**(b)**	#8 copper	551.56(C)
28.	**(c)**	140°F	427.12
29.	**(c)**	80%	501.130(B3)
30.	**(c)**	no external obst.	600.9(D1,2,3)
31.	**(c)**	door	490.44(B)
32.	**(a)**	grounded	427.29
33.	**(d)**	both b & c	760.2 DEF
34.	**(c)**	sufficient pressure	506.2 DEF
35.	**(d)**	25 amperes	411.2
36.	**(b)**	inside or outside	408.4
37.	**(d)**	6"	352.20(B)
38.	**(d)**	CI	760.2 DEF
39.	**(b)**	4"	110.31(A)
40.	**(a)**	less	215.2(B1)
41.	**(c)**	86°F	324.10(F)
42.	**(b)**	1/2"	344.20(A)
43.	**(d)**	CATVP cable	820.154(A)
44.	**(c)**	15v	680.23(A3)
45.	**(a)**	ClassI, Gp. D, Div.2	T.514.3(B1)
46.	**(d)**	all receptacles	511.12
47.	**(c)**	20'	516.3(C1)
48.	**(c)**	at both	514.1 514.9(A)
49.	**(d)**	all of these	517.33(A1, 8d&g)
50.	**(c)**	50% or both	210.23(A2)
51.	**(d)**	grd. electrode	230.42(C) & 250.24(B)
52.	**(a)**	additional rod	250.56
53.	**(a)**	metal raceways	300.22(C1)
54.	**(d)**	hard service	680.56(B)
55.	**(d)**	prevent	700.6(A)
56.	**(c)**	2'	T.300.50
57.	**(c)**	7'	760.30(A1)
58.	**(a)**	equivalent object.	90.4
59.	**(b)**	6'6" = 6 1/2'	110.26(E)
60.	**(a)**	any terminal	200.11
61.	**(a)**	humidifier	422.12 ex.
62.	**(d)**	same raceway	408.20
63.	**(d)**	substitute	400.8(1)
64.	**(b)**	12" = 1'	410.67(C)
65.	**(a)**	NEMA Type 1	T.430.91
66.	**(d)**	all of the above	430.53(A1,2,3)
67.	**(c)**	incandescent	410.8(B1,2)
68.	**(c)**	115%	445.13
69.	**(d)**	all of the above	440.62(A1,2,3)
70.	**(b)**	18"	513.10(C1)

ANSWERS MASTER OPEN BOOK EXAM #2

1. **(c)**	#6	800.100(D)	36. **(c)**	two hours	700.12(B2)
2. **(d)**	200%	630.12(A)	37. **(d)**	I,II or III	300.50(A2)
3. **(c)**	next higher	430.52(C1) ex.1	38. **(c)**	accessible only to	700.25
4. **(c)**	externally wired	410.7	39. **(b)**	Proscenium	520.2 DEF
5. **(b)**	20 amps	600.5(B1)	40. **(c)**	I & II only	604.6(C)
6. **(d)**	unbroken lengths	386.10(4)	41. **(c)**	liquidtight flex	553.7(B)
7. **(a)**	Class I, Div.1	517.60(A2)	42. **(a)**	rms	620.13 FPN1
8. **(a)**	general-purpose	501.115(B2)	43. **(d)**	shall not	605.5(C)
9. **(a)**	1/4"	600.41(C)	44. **(d)**	water cooled	503.125
10. **(d)**	all of these	620.11(A)	45. **(b)**	10 amperes	725.23
11. **(a)**	18"	501.15(A1)	46. **(c)**	firestopped	760.3(A) 300.21
12. **(c)**	600 amperes	T.392.7(B) note b	47. **(c)**	carrier	770.9(A,B,C)
13. **(d)**	I,II,III	517.2 DEF FPN	48. **(d)**	concrete	800.47(A)
14. **(a)**	gases & vapors	500.5(B)	49. **(d)**	#14 cu	800.100(A3)
15. **(d)**	48 times	314.71(A)	50. **(b)**	Class I, Div.2	500.5(B2)(1)
16. **(b)**	simultaneously	514.11(A)	51. **(a)**	Class 1	725.11(A)
17. **(d)**	both a & b	700.12(A)	52. **(c)**	MI cable	300.22(B)
18. **(d)**	Class I, Div.1	511.3(B3)(2)	53. **(d)**	50%	T.310.15(B2a)
19. **(c)**	polarity	200.11	54. **(d)**	600v	725.21(B)
20. **(a)**	rigid metal	501.10(A1)	55. **(b)**	bare or insulated	250.118(1)
21. **(b)**	excluding moisture	501.15	56. **(a)**	25%	501.15(C6)
22. **(b)**	100 or more	518.1	57. **(d)**	30v	725.21(A)
23. **(d)**	metal hood	520.24	58. **(b)**	made or broken	513.4(A)
24. **(d)**	I,II,III	600.9(D1,2,3)	59. **(b)**	GFCI	640.10(A)
25. **(d)**	not allowed	502.100(A3)	60. **(d)**	all of these	760.8
26. **(b)**	#6 copper	280.23	61. **(b)**	external	505.25(B)
27. **(a)**	400 sq.ft.	552.2 DEF	62. **(d)**	125%	645.5(A)
28. **(c)**	10v	517.64(A1)	63. **(d)**	single means	620.51(C1)
29. **(c)**	intrinsically safe	504.80(C)	64. **(c)**	18"	760.30(A1)
30. **(a)**	adequate bonding	250.116 FPN	65. **(b)**	protector	800.90(B) FPN
31. **(b)**	II only	725.55(A)	66. **(c)**	flexible metal	725.55(I)
32. **(d)**	I,II,III	669.9(1,2,3)	67. **(b)**	Class II	502.115(A2)
33. **(b)**	equip. grd.	515.8(C)	68. **(d)**	I,II or III	800.100(B1)(1,4,5)
34. **(a)**	electrolytic cell	668.2 DEF	69. **(a)**	7 amp, 10 amp	760.23
35. **(c)**	funct.associated	725.26(B1)	70. **(b)**	designed	430.88 ex.

ANSWERS MASTER OPEN BOOK EXAM #3

1. **(c)**	18"	501.15(A1)
2. **(d)**	any of the above	515.8(A)
3. **(c)**	energized	520.73
4. **(c)**	Orange-Brn.	517.61(A1) 517.160(A5)
5. **(a)**	#2 bare copper	250.52(A4)
6. **(c)**	NEMA Type 6	T.430.91
7. **(c)**	sprinkler	450.47
8. **(d)**	6' with barbed wire	460.2(A 110.31
9. **(b)**	300%	T.450.3(A)
10. **(a)**	Y connection	430.7(B2)
11. **(c)**	35 pounds	314.27(D)
12. **(d)**	fixture & ballast	410.73(F1,4)
13. **(c)**	#8 copper	400.31(A)
14. **(b)**	50%	404.14(B2)
15. **(c)**	lockout "off"	427.56(D1,2,3)
16. **(c)**	150%	430.22(A) ex.B
17. **(a)**	50%	430.22(D)
18. **(b)**	wired	640.21(D1,2,3,4)
19. **(a)**	3'	516.3(b4)
20. **(a)**	left side	551.77(A)
21. **(c)**	10'	516.3(C2b)
22. **(a)**	charged	513.10(A2)
23. **(a)**	handle	511.4(B2)
24. **(c)**	grounded	504.50(C)
25. **(a)**	12"	501.130(B3)
26. **(b)**	enclosure	502.115(A1)
27. **(a)**	terminal & junction	760.10
28. **(b)**	3 hours	110.31(A)
29. **(c)**	impedance	427.2 DEF
30. **(d)**	125%	210.19(B1)
31. **(c)**	insulating material	410.42(B1,2,3)
32. **(b)**	thermal relays	240.9
33. **(b)**	two breakers	408.36(A)
34. **(b)**	voltage	406.6
35. **(b)**	corrosion	312.10(A)
36. **(c)**	three	240.100(1)
37. **(d)**	36 times	314.71(B1)
38. **(b)**	100v	250.188(C)
39. **(a)**	six times	336.24(3)
40. **(b)**	voltage	404.15(A)
41. **(a)**	off position	426.51(B)
42. **(a)**	should be sealed	300.7
43. **(d)**	2000v	310.7
44. **(c)**	both a & b	392.6(F1,2)
45. **(d)**	2"	398.19
46. **(d)**	wet or corrosive	340.10(3)
47. **(c)**	support of fixtures	352.12(B)
48. **(d)**	12"	350.30(A)
49. **(d)**	flush or project	314.20
50. **(d)**	167%	725.21(A2)
51. **(d)**	no separate	610.42(B3)
52. **(b)**	15 minutes	700.12(B1)
53. **(b)**	#14 cu	600.7(D)
54. **(a)**	Class 1	725.11
55. **(c)**	60 amperes	680.9
56. **(c)**	connect equip. grd.	680.26(C1,2,3)
57. **(c)**	10 amperes	725.23
58. **(a)**	emergency system	700.6(A)
59. **(a)**	integrated	427.2 DEF
60. **(a)**	115%	430.128
61. **(b)**	GFCI	620.85
62. **(b)**	125%	430.122(A)
63. **(d)**	disconnect	430.84 ex.
64. **(b)**	looping cables	620.41(1,2,3)
65. **(c)**	3' x 2'	600.21(E)
66. **(a)**	one duplex	620.24(C)
67. **(c)**	stranded	610.11(C)
68. **(d)**	30 ampere	551.71
69. **(a)**	fire protection	450.47
70. **(d)**	2'	504.30(A3)

ANSWERS MASTER OPEN BOOK EXAM #4

1. **(d)**	motor O.L.	430.224
2. **(c)**	23 amps	Table 402.5
3. **(a)**	120 watts	426.20(A)
4. **(b)**	20 amps	520.9
5. **(c)**	I & II	470.18(B)
6. **(b)**	approved	426.21(A)
7. **(c)**	2"	501.115(B1)(2)
8. **(d)**	limiting means	422.47
9. **(a)**	150%	430.6(C)
10. **(a)**	two branch circuits	517.18(A)
11. **(c)**	13.6a	Table 430.147
12. **(c)**	#8 insulated	680.27(A2)
13. **(c)**	adequate	450.9
14. **(d)**	6-phase wye	647.5
15. **(a)**	capacitor	455.2 DEF
16. **(d)**	36 1/2'	552.44(B)
17. **(a)**	one duplex	620.23(C)
18. **(d)**	50 volts	460.28(A)
19. **(d)**	all of these	455.8(B)
20. **(b)**	25'	625.17
21. **(c)**	organic	500.8(C2)
22. **(c)**	13 receptacles	605.8(C)
23. **(c)**	portable	501.140(A4)
24. **(c)**	6"	426.23(A)
25. **(c)**	#14 - #2	334.104
26. **(b)**	exposed indoor	354.12(1)
27. **(d)**	I,II,III & IV	501.105(B6)
28. **(d)**	20'	514.11(C)
29. **(c)**	adjustable	430.52(C3)
30. **(b)**	ungrounded	590.4(D)
31. **(c)**	dusttight	503.10(A1)
32. **(c)**	III only	680.43(D) ex.
33. **(c)**	8'	517.63(C1)
34. **(b)**	two crossings	324.18
35. **(a)**	I only	110.14(C)
36. **(a)**	25 amps 50a x 50% =25a	517.73(A)
37. **(c)**	means provided	620.91(A)
38. **(b)**	2"	520.47
39. **(a)**	step & touch volt	682.23
40. **(b)**	#18	660.9
41. **(a)**	reinforced 6" thick	450.42 FPN2
42. **(c)**	closed construction	545.2 DEF
43. **(d)**	#14 copper	675.10(A3)
44. **(d)**	patient grd. point	517.2 DEF
45. **(d)**	integrated system	685.1
46. **(b)**	manufacturer's	430.52(C2)
47. **(a)**	4'	625.29(B)
48. **(d)**	B phase	408.3(E)
49. **(d)**	all of these	605.8(A)
50. **(a)**	barriers	430.97(E)
51. **(a)**	ground fault protection	690.5
52. **(d)**	100%	T. 530.19(A)
53. **(c)**	either I or II	422.32
54. **(c)**	12'	370.6(A)
55. **(c)**	28 amps	Table 520.44
56. **(c)**	general purpose	668.40
57. **(d)**	III & IV	366.2
58. **(a)**	vapor seal	368.234(A)
59. **(a)**	one ampere	780.5
60. **(c)**	12"	820.44(B) ex.
61. **(b)**	3 hours	450.42
62. **(d)**	I,II & III	760.1 FPN1
63. **(a)**	I only	552.4
64. **(c)**	I & II only	800.113 FPN1
65. **(c)**	#14 copper	810.58(C)
66. **(a)**	black	322.120(C)
67. **(a)**	Article 518	518.2
68. **(d)**	low water cut off	680.51(D)
69. **(c)**	entire b.c.	210.12(B)
70. **(d)**	80' - 50'	610.21(D)

ANSWERS MASTER OPEN BOOK EXAM #5

1.	**(a)**	not have a pole	690.13		
2.	**(c)**	flexible metal	725.55(I)		
3.	**(b)**	6"	520.53(H2)		
4.	**(d)**	100 pounds	110.75(D)		
5.	**(b)**	separate b.c.	620.22(A)		
6.	**(d)**	separated 0.25"	760.55(D1)		
7.	**(c)**	service conductors	DEF 100		
8.	**(b)**	may be installed	DEF 100 • 240.24(E)		
9.	**(d)**	GFCI required	210.52(G) • 210.8(A5)		
10.	**(d)**	all of these	440.6(A)		
11.	**(c)**	FLC x 125%	455.7(A)		
12.	**(c)**	150v to grd	810.16(B)		
13.	**(d)**	facilitate O.C.D.	250.4(B4)		
14.	**(c)**	snap switch	430.109(C2)		
15.	**(d)**	3' from door	225.19(D1)		
16.	**(c)**	90°C	320.80(A)		
17.	**(a)**	single, no GFCI	620.85		
18.	**(d)**	shall be grd.	406.5(B)		
19.	**(a)**	not permitted	230.70(A2)		
20.	**(d)**	GFCI breaker	406.3(D3c)		
21.	**(a)**	90°C	410.33		
22.	**(d)**	25'	410.77(C)		
23.	**(a)**	two screws	406.4(C)		
24.	**(a)**	require GFCI	680.44		
25.	**(d)**	all of these	400.10 FPN		
26.	**(a)**	hinged base	410.15(B) ex.2		
27.	**(c)**	limited care facility	517.2 DEF		
28.	**(d)**	inside or outside	250.102(E)		
29.	**(c)**	GFCI required	406.3(D2)		
30.	**(c)**	18"	422.16(B1)(2)		
31.	**(d)**	disconnect grd.	514.11(A)		
32.	**(b)**	threaded hubs	314.23(E)		
33.	**(d)**	60 amps	Table 250.122		
34.	**(b)**	#4	460.8 & T.310.16		
35.	**(a)**	.25" = 1/4"	352.44		
36.	**(a)**	GFCI protected	590.6(A)		
37.	**(b)**	3'	408.18(A)		
38.	**(b)**	enclosed raceway	250.64(B)		
39.	**(d)**	equip. grd. cond.	DEF 100		
40.	**(b)**	1 1/4"	300.4(A1)		
41.	**(b)**	#12	240.5(B2)		
42.	**(a)**	#8	250.36(B)		
43.	**(a)**	factory installed	600.10(C2)		
44.	**(d)**	expansion & cont.	352.30		
45.	**(b)**	fully selective GFCI	517.17(C)		
46.	**(c)**	must have GFCI	210.8(A3)		
47.	**(b)**	16,510cm	Table 8		
48.	**(c)**	75%	388.56		
49.	**(b)**	50%	Table 310.15(B2a)		
50.	**(d)**	5%	210.19 • 215.2		
51.	**(a)**	must have GFCI	210.8(B1)		
52.	**(a)**	dry walls	388.10(2)		
53.	**(d)**	#1/0 copper	Table 250.66		
54.	**(b)**	FPN	90-5(C)		
55.	**(d)**	#2/0 copper	Table 250.66		
56.	**(b)**	12"	820.44(B) ex.		
57.	**(a)**	1.5 hp	Table 430.148		
58.	**(b)**	7'	110.31		
59.	**(c)**	GFCI not required	210.8(A2) ex.2		
60.	**(b)**	6'	250.102(E)		
61.	**(a)**	25%	430.24		
62.	**(d)**	controller & discon.	430.102(A) & 103		
63.	**(a)**	#18	330.104		
64.	**(a)**	#6 copper	250.66(A)		
65.	**(b)**	nonincendive	DEF 100		
66.	**(d)**	equip. enclosure	250.96(B)		
67.	**(d)**	6' above floor	410.101(C8)		
68.	**(d)**	50 pounds	314.27(B)		
69.	**(a)**	raceways	725.58		
70.	**(d)**	white or gray	200.6(A)		

ANSWERS MASTER OPEN BOOK EXAM #6

1.	**(d)**	150v - 1000v	665.22	36.	**(b)**	II only	480.10(B)
2.	**(b)**	1/2" HDPE	353.20(A)	37.	**(b)**	50 kva	450.45(C)
3.	**(c)**	I,II or IV	760.71(H)	38.	**(c)**	9'	605.5(B)
4.	**(c)**	17°C	310.10 FPN2	39.	**(b)**	intermittent	620.61(B4)
5.	**(c)**	I,III & IV	516.10(A8)(1,2,3)	40.	**(a)**	limited care facility	517.2 DEF
6.	**(a)**	10 amps	552.52(A)	41.	**(b)**	60%	392.11(B2)
7.	**(a)**	1000va	725.21(A)	42.	**(b)**	II only	322.10(1)
8.	**(c)**	35,000v	450.23(A1a)	43.	**(d)**	close together	408.56
9.	**(d)**	coordinated	700.27	44.	**(c)**	grounded	660.47
10.	**(a)**	I only	630.41	45.	**(d)**	25'	410.77(C)
11.	**(c)**	30 amps	660.5	46.	**(b)**		680.27(C1,2)
12.	**(c)**	life safety branch	517.2 DEF	47.	**(c)**	suitable for ac	430.109(C2)
13.	**(b)**	Class 2	725.41(A5)	48.	**(d)**	I,III, & IV	520.5
14.	**(c)**	normally closed	520.49	49.	**(a)**	8 conductors	312.11(C)
15.	**(a)**	NPLFA	760.2	50.	**(d)**	70%	551.71
16.	**(d)**	closed loop	780.1	51.	**(c)**	I,II and IV	430.245(B)
17.	**(a)**	I & II only	404.13(B)	52.	**(a)**	I only	520.6
18.	**(a)**	1"	300.39	53.	**(a)**	20' - 100'	514.11(C)
19.	**(d)**	150%	430.6(C)	54.	**(d)**	I,II & III	725.11(A)
20.	**(d)**	30'	605.8	55.	**(c)**	I & II only	680.33(A3,4)
21.	**(c)**	60 volts	820.15	56.	**(d)**	#24 copper	780.6(A)
22.	**(a)**	10 volts	517.64(F) FPN	57.	**(d)**	#10 copper	810.58(B)
23.	**(c)**	I & III only	518.5	58.	**(b)**	each ungrounded	285.4
24.	**(c)**	IV only	550.2 DEF	59.	**(d)**	300 volts	250.170
25.	**(d)**	600 volts	620.3(C)	60.	**(d)**	0.77	Table 620.14
26.	**(d)**	I or III only	695.6(A,B)	61.	**(d)**	I,II or III	668.32(B)
27.	**(b)**	24 volts	517.63(D)	62.	**(d)**	I,II & III	517.17(B1,2,3)
28.	**(c)**	II or III only	610.32	63.	**(b)**	over 250 volts	440.60
29.	**(d)**	300%	630.32(B)	64.	**(d)**	20'	250.188(E)
30.	**(d)**	6 times	440.12(C)	65.	**(c)**	15'	398.30(B)
31.	**(d)**	#6 copper	820.100(D)	66.	**(b)**	2"	800.133(A2)
32.	**(c)**	20 ampere	660.9	67.	**(a)**	20%	660.6(B)
33.	**(c)**	III only	700.21	68.	**(a)**	60%	675.22(A)
34.	**(a)**	50 volts	690.41	69.	**(b)**	3 interconnections	520.53(J)
35.	**(b)**	stand-alone sys.	692.2	70.	**(b)**	37 CFM	Table 625.29(D2)

ANSWERS MASTER OPEN BOOK EXAM #7

1. **(c)** 1/2"	410.46	36. **(d)** I,II & III	530.41
2. **(b)** NPT	500.8(D)	37. **(d)** not classified	511.3(A)
3. **(b)** 6'	210.50(C)	38. **(b)** 6 pounds	410.15(A)
4. **(d)** cabinet	DEF 100	39. **(c)** solar cell	690.2 DEF
5. **(d)** 25'	250.86(2)	40. **(a)**	90.2(A1)
6. **(c)** both	230.54(F)	41. **(a)** 7'	411.5(C)
7. **(b)** 0.027"	Chapter 9 Table 8	42. **(c)** 12"	410.8(A)
8. **(c)** 75a	240.6	43. **(a)** 10'	225.4
9. **(a)** fork lift	625.2 DEF	44. **(d)** all	400.10 FPN
10. **(c)** II, III and IV	630.1	45. **(d)** either	230.22
11. **(c)** carpet squares	324.1	46. **(c)** #10	517.14
12. **(c)** 2 1/2' = 30"'	230.51(A)	47. **(a)** 12"	Chap.9 Table 4
13. **(b)** can be started	430.7(B1)	48. **(a)** 3"	424.41(D)
14. **(a)** 35'	810.11 ex.	49. **(d)** I, II & III	324.40(A)
15. **(a)** red	424.35	50. **(d)** 150v	520.25(C)
16. **(c)** 42"	680.2 DEF	51. **(a)** 50 amps	680.26(E)
17. **(c)** I, II & IV	T.310-13	52. **(d)** I, II & III	511.7(A1)
18. **(d)** 15'	525.5(B)	53. **(d)** unbroken lengths	386.10(4)
19. **(c)** #18 or larger	240.5(B1)	54. **(b)** 1ø - 3ø	240.85
20. **(c)** 50a	422.11(C)	55. **(c)** freedom from hazard	90.1(B)
21. **(c)** end seal	332.80	56. **(b)** 5'	368.30
22. **(b)** clothes closets	240.24(D)	57. **(c)** III only	680.26(C)
23. **(b)** 48"	230.54 ex. = 2 x 24" = 48"	58. **(d)** inductive current	300.20(A)
24. **(a)** 1/2" to 4"	350.20(A,B)	59. **(a)** 50w	422.43(A)
25. **(d)** I,II,III & IV	630.31(B) FPN	60. **(c)** free circulation	110.13(B)
26. **(d)** 200a	404.16	61. **(d)** duty cycle	409.20
27. **(c)** 18'	210.6(D1b)	62. **(d)** 100'	240.21(B4)(2)
28. **(a)** tamper resistant	517.18(C)	63. **(d)** insulated	550.16(A1)
29. **(b)** 50a	310.15(B2) ex.3	64. **(d)** GFCI outlet	422.51
30. **(b)** MTW	Table 310.13	65. **(c)** 10"	Table 408.5
31. **(d)** EMT	502.10(A1)	66. **(d)** #12	680.23(F2)
32. **(b)** 10 seconds	700.12	67. **(a)** 600v	250.184(A)
33. **(d)** I, II & III	250.64(A)	68. **(b)** 167°F	Table 402.3
34. **(a)** RMS	620.13 FPN	69. **(c)** I & III only	517.30(B1)
35. **(c)** flame arrestor	480.10(A)	70. **(a)** 15a	408.52

ANSWERS MASTER OPEN BOOK EXAM #8

1.	**(d)** I,II & III	410.16(C)
2.	**(a)** MTW	310.8(C2)
3.	**(c)** disconnect all	210.4(B)
4.	**(c)** individual	450.2
5.	**(c)** afford protection	110.12(A)
6.	**(d)** maintained	310.15(B2b)
7.	**(c)** liquidtight	553.7(B)
8.	**(a)** 600v	225.10
9.	**(c)** both I & II	300.20(B)
10.	**(c)** 50%	Table 220.44
11.	**(c)** authority	555.13(B3)
12.	**(c)** 1/3	450.5 FPN
13.	**(b)** identified	780.3(A)
14.	**(c)** underground	310.60(A)
15.	**(d)** indicate	110.22
16.	**(c)** major diameter	Chap. 9 Table 1 note 9
17.	**(b)** II only	700.26
18.	**(a)** twice	220.55
19.	**(d)** MV	328.2 DEF
20.	**(d)** all	430.32(E)
21.	**(c)** 250%	450.6(B)
22.	**(c)** #8	540.13
23.	**(d)** 20'	600.32(J1)(1)
24.	**(a)** 20a	406.2(C)
25.	**(a)** outlet	600.5(A)
26.	**(a)** 300 ma	600.23(D)
27.	**(b)** II only	410.76(A)
28.	**(b)** intrinsically safe	504.2 DEF
29.	**(b)** suitable	110.8
30.	**(d)** flex. nonmetallic	394.19(B)
31.	**(d)** I,II & III	90.2(A)1,2
32.	**(d)** lighting & power	110.19
33.	**(b)** one in ungrd.	Table 430.37
34.	**(c)** I & II only	300.15(A)
35.	**(a)** 2 feet	555.2(2)
36.	**(b)** I & II only	700.15
37.	**(b)** .1194	Chapter 9 Table 5(A)
38.	**(b)** 20a	422.11(E2)
39.	**(d)** 87 1/2%	701.11(A)
40.	**(b)** 1 second	230.95(A)
41.	**(d)** loop connectors	300.19(B1,2,3)
42.	**(b)** car lights	620.22(A)
43.	**(b)** 12'	Table 300.5
44.	**(a)** installed	230.95(C)
45.	**(b)** insulating mats	430.233
46.	**(d)** high-voltage	314.72(E)
47.	**(c)** III only	392.8(E)
48.	**(c)** 6 times	230.208
49.	**(b)** hoistways	386.12(4)
50.	**(b)** either side	501.15(B2)
51.	**(b)** 3/4"	390.3(A)
52.	**(d)** 6 1/2"	T. 312.6(B) note 2
53.	**(c)** #14	820.100(A3)
54.	**(c)** 8' vert. & 5' hor.	250.110 (1)
55.	**(b)** raintight	300.6(A)
56.	**(d)** I,II & III	406.8(A)
57.	**(b)** weatherproof	312.2(A)
58.	**(b)** I, II & IV	362.10
59.	**(a)** 100%	450.6(A2)
60.	**(c)** 125%	422.10(A)
61.	**(c)** simultaneously	230.74
62.	**(b)** 2 1/2"	Table 312.6(A)
63.	**(d)** 150%	445.12(C)
64.	**(d)** I, II & III	332.10(1,3,5)
65.	**(d)** uninsulated	338.100
66.	**(d)** 6 times	430.110(C3)
67.	**(c)** shielded	240.41(A)
68.	**(d)** 24'	310.11(B1)
69.	**(d)** 15'	600.10(D2)
70.	**(d)** I or II only	440.6(A) ex.

ANSWERS MASTER OPEN BOOK EXAM #9

1. **(d)** I,II or III	250.20(A1,2,3)
2. **(c)** both I & II	250.34(A1,2)
3. **(c)** 120/240 1 ø	550.30
4. **(d)** I,II & III	280.4B FPN2
5. **(b)** listed	240.20C
6. **(d)** cable tray	392.2 DEF
7. **(a)** offset	312.6(B2) FPN
8. **(b)** box or fitting	314.27(B)
9. **(c)** both I & II	314.25 • 314.28(B)
10. **(d)** neither I nor II	551.20(E)
11. **(d)** same raceway	310.4 (1)(2)(5)
12. **(a)** 5 times	332.24(1)
13. **(a)** MI	300.37
14. **(d)** I,II & III	760.21, 23, 26(A)
15. **(a)** #8	547.10(B)
16. **(b)** single	427.47
17. **(d)** as far away as possible	450.45(A)
18. **(b)** insulated	394.12(5)
19. **(c)** GFCI	422.49
20. **(b)** 8 times	300.34
21. **(d)** 60 volts	647.1
22. **(c)** 60°C	424.11
23. **(b)** 80%	501.8(A4)
24. **(a)** #18	410.27(B)
25. **(a)** readily identified	700.9(A)
26. **(a)** external means for adjusting	240.6(B)
27. **(b)** appliance	422.14
28. **(b)** electrostatic	516.10(B4)
29. **(c)** I,III & IV	300.22(B)
30. **(a)** 10 kva	517.160 FPN1
31. **(c)** written record	700.4(D)
32. **(c)** both I & II	230.81
33. **(d)** I, II & III	410.29(A,B,C1)
34. **(a)** distinctive	200.6(A1)
35. **(d)** I, II, & III	700.12(B1,2,3)
36. **(b)** braces or guys	230.28
37. **(a)** 10'	225.18
38. **(d)** I, II & III	680.6(2,5,6)
39. **(b)** 5"	Table 430.10(B)
40. **(a)** internal	90.7
41. **(c)** continuous duty	T. 430.22(E) note
42. **(a)** ungrounded	240.20(A)
43. **(c)** II and III only	422.16(B4)(3,5)
44. **(d)** 100'	620.41 (2)
45. **(d)** own section	702
46. **(b)** 90°	110.26(A2)
47. **(d)** I, II & III	640.1
48. **(b)** II only melting point	240.100(B)
49. **(d)** not required	250.162(A)
50. **(c)** permanent	517.2 DEF
51. **(d)** anesthetizing	517.160(B3) FPN
52. **(d)** I, II or III	450.25
53. **(d)** rack supported	695.12(C)
54. **(c)** 300v	410.73(B)
55. **(c)** I & III only	701.2 FPN
56. **(d)** need **not**	250.172 ex.
57. **(d)** rigid	344.10(B)
58. **(b)** metal	372.6
59. **(b)** II only	424.29
60. **(c)** I and II only	727.4 (1)(7)
61. **(b)** 3'	225.19(B)
62. **(c)** 0.829	Table 4
63. **(c)** 12 1/2"	Table 360.24(A)
64. **(b)** I & II only	240.83(E)
65. **(c)** both I and II	725.2 DEF
66. **(b)** I, II & IV	430.242(1,2,4)
67. **(d)** I, II or III	690.2 DEF
68. **(c)** 3.5 kva	Annex D Example D2b
69. **(c)** NFPA	90.6
70. **(d)** 1.342	Chapter 9 Table 4

ANSWERS MASTER OPEN BOOK EXAM #10

1. **(a)** 8"	314.17(C) ex.	
2. **(d)** all of the above	230.30 ex.	
3. **(d)** I, II or III	800.100(B1)	
4. **(a)** 6 1/2'	210.52(E)	
5. **(d)** I,II or III	250.102(A)	
6. **(d)** 12"	682.15	
7. **(d)** 30 amps	530.14	
8. **(c)** bottom shield	324.2 DEF	
9. **(a)** comb. dust	500.5(C)	
10. **(b)** 200 amps	110.26(E) ex.	
11. **(c)** both	675.15 • 250.53B	
12. **(c)** I or II	422.40	
13. **(c)** dist. marked	250.110 ex.3 • 250.114 ex.	
14. **(d)** 36.96	T. 610.14(A)	33a x 112%
15. **(a)** grd. cond.	200.3	
16. **(c)** I & II only	604.6(C)	
17. **(c)** 6 times	430.110(C3)	
18. **(a)** recpt. outlet	210.50(A)	
19. **(a)** 1/2"	520.43(A)	
20. **(b)** 400%	530.18(B)	
21. **(d)** I,II,III & IV	280.21	
22. **(b)** splices & taps	352.56	
23. **(d)** I, II or III	110.14(A)	
24. **(d)** I, II & III	334.40(B)	
25. **(a)** 4	517.2 DEF	
26. **(b)** fountains	680.51(A)	
27. **(b)** 300%	630.32(A)	
28. **(a)** ungrounded	404.2(A)	
29. **(a)** #6	230.202(A)	
30. **(d)** csa 45%	374.5	
31. **(d)** I,II, or III	300.50(A2)	
32. **(c)** 125%	690.8(B1)	
33. **(b)** 20a or less	250.118(6b)	
34. **(b)** grounded	200.10(C)	
35. **(b)** II only	550.10(A)	
36. **(c)**	250.118 (6b)	
37. **(d)** 16	T. 344.30(B2)	
38. **(d)** none	250.64(C)	
39. **(a)** 3/4"	Table 408.56	
40. **(b)** II only	440.2 DEF	
41. **(d)** 24"	Chapter 9 note 4	
42. **(c)** #6	250.66(A)	
43. **(b)** readily accessible	240.10	
44. **(c)** 2.15	392.11(B4)	
45. **(a)** I & II only	210.8(A1,4)	
46. **(b)** 48 ... 60	680.9	
47. **(c)** I or II	460.25(C)	
48. **(a)** 35 lb.	314.27(D)	
49. **(c)** 12 1/2%	250.24(C1)	
50. **(d)** 1 1/2"	410.76(B)	
51. **(a)** 25'	210.63	
52. **(b)** 1/8"	410.50	
53. **(d)** larger kw demand	T. 555.12 note 1	
54. **(b)** 511 • 517	510.1	
55. **(d)** source of supply	230.204(D)	
56. **(b)** 6"	404.5 ex.	
57. **(c)** 50'	368.17(B) ex.	
58. **(d)** header	372.9	
59. **(c)** both I & II	240.82	
60. **(d)** water cooled	503.125	
61. **(d)** 150 amps	424.72(A)	
62. **(c)** both I & II	430.35(A,B) ex.	
63. **(d)** 125%	424.82	
64. **(c)** both I & II	250.24(C1)	
65. **(c)** 300v	680.51(B)	
66. **(c)** both I & II	513.3(A,C)	
67. **(b)** louvered	450.43A,B • 450.45(D)	
68. **(a)** trolley frame	610.61	
69. **(a)** short sections	300.12 ex.	
70. **(b)** 30 '	Table 300.50	

ANSWERS MASTER OPEN BOOK EXAM #11

1.	**(b)** #20	620.12(A1)	36.	**(c)** bonding jumper	250.98
2.	**(a)** insulated	300.3(C1)	37.	**(b)** ex. moisture	501.15
3.	**(d)** 35kv	450.21(C)	38.	**(a)** 8	408.35
4.	**(a)** current-carrying	520.27(B)	39.	**(c)** 3"	810.54
5.	**(c)** .040	410.38(A)	40.	**(a)** 6"	Table 300.5
6.	**(d)** 1/4" = .25"	378.44	41.	**(b)** simult.	514.11(A)
7.	**(a)** 8'	Table 400.4 note 2	42.	**(b)** 18"	210.52(A3)
8.	**(b)** outside	220.12	43.	**(a)** rigid metal	501.10(A1a)
9.	**(d)** I, II & III	110.18	44.	**(b)** #10	545.4(B)
10.	**(b)** Class II, Div.I	500.5(C1)(1)	45.	**(d)** Class I, Div.I	511.3(B3)(3)
11.	**(b)** 25a	210.3	46.	**(d)** I, II & III	250.10(2)
12.	**(d)** #14	620.12(A1)	47.	**(d)** 3 hours	450.43(A)
13.	**(c)** 600a	T.392.7(B2) **note	48.	**(c)** source of heat	410.54(A)
14.	**(c)** III only	250.4(B4)	49.	**(c)** 20a	210.19(A3) ex.
15.	**(c)** polarity	200.11	50.	**(b)** 8.19a	T. 630.31A2 • 21a x .39
16.	**(c)** not more than 6"	110.26(A3)	51.	**(d)** all	310.11(A1,2,3)
17.	**(a)** 1va	220.14(K2)	52.	**(d)** I,II or III	250.30(A1)
18.	**(a)** at all	422.11(A)	53.	**(c)** similar change	240.23
19.	**(c)** I & II	110.11	54.	**(a)** gases & vapors	500.5(B)
20.	**(d)** metal hood	520.24	55.	**(b)** 15a	322.56(B)
21.	**(a)** 3"	408.5	56.	**(c)** phase converter	455.2 DEF
22.	**(c)** I, II & III	230.50(A1,2,4)	57.	**(d)** #8	230.31(B)
23.	**(b)** 100	518.1	58.	**(a)** 18"	501.15(A1)
24.	**(b)** #4	250.64(B)	59.	**(c)** 3/4"	344.28
25.	**(d)** both	700.12(A)	60.	**(b)** grounding cond.	DEF 100
26.	**(a)** D	310.11(C)	61.	**(d)** 6' 7"	404.8(A)
27.	**(b)** II only	392.11(A1)	62.	**(a)** fault levels	110.9
28.	**(c)** pigtail	300.13(B)	63.	**(b)** will not	230.95 FPN1
29.	**(d)** grd. electrode	DEF 100	64.	**(a)** general-use type	501.115(B1)(2)
30.	**(b)** beyond outside walls	300.5(C)	65.	**(c)** both	334.15(B) 334.23
31.	**(d)** V	334.116(B)	66.	**(d)** Class I, Group C	517.61(A5)
32.	**(a)** will not burn	240.41(A)	67.	**(b)** 2'	320.30(B2)
33.	**(d)** I, II & III	517.2 DEF FPN	68.	**(a)** Class I, Div.I	517.60(A1)
34.	**(b)** unbalanced	220.61(A)	69.	**(b)** 30a	600.5(B2)
35.	**(d)** 48 times	314.71(A)	70.	**(a)** identified	230.28

ANSWERS MASTER OPEN BOOK EXAM #12

1. **(b)** same size	250.122(A)	
2. **(b)** 10°	310.13 FPN • 402.3 FPN	
3. **(a)** raintight	230.54(A)	
4. **(c)** two or more	230.40 ex.1	
5. **(c)** 18.5 feet	T. 225.60	
6. **(c)** right angle	372.5	
7. **(d)** 2"	501.15(A1)(2)	
8. **(a)** listed	600.3	
9. **(c)** 400%	530.18(B)	
10. **(a)** 3'	230.9 ex.	
11. **(c)** 20 cells	480.6(B)	
12. **(b)** within sight	600.6(A1)	
13. **(c)** temperature rise	470.18(E)	
14. **(c)** 462 amps	T. 555.12	
15. **(c)** 40a	440.62(A2)	
16. **(d)** listing	110.3(B)	
17. **(a)** I only	210.52B1	
18. **(b)** 300v	Table 402.3	
19. **(d)** #10	225.6(A1)	
20. **(a)** 5'	680.22(A1)	
21. **(d)** 20'	250.52(A3)	
22. **(c)** flex.metal	501.10(B2)(2)	
23. **(a)** 4 1/2'	348.30(A)	
24. **(c)** 135%	460.8(A)	
25. **(d)** I, II & III	645.5(D1,3,4)	
26. **(d)** adjacent	310.15(B3)	
27. **(b)** 90°	410.5 • 410.65(A)	
28. **(d)** elect. cont.	250.64(E)	
29. **(d)** not classified	513.3(D)	
30. **(d)** water heater	550.2 FPN	
31. **(a)** trans. assembly	324.40(D)	
32. **(d)** I, II & III	700.7(A,B,C)	
33. **(d)**	330.10(B4)	
34. **(b)** 2 1/2' = 30"	250.53(F)	
35. **(d)** six	517.19(B)	
36. **(a)** 25%	430.24	
37. **(b)** identified	344.120	
38. **(c)** II & III	250.62	
39. **(c)** 15-20-30	424.3(A)	
40. **(b)** continuity	392.6(A)	
41. **(d)** I, II & III	545.8	
42. **(c)** therm. protector	DEF 100	
43. **(c)** 8'	410.4(D)	
44. **(c)** hazard	240.4(A)	
45. **(c)** #10	110.14(A)	
46. **(a)** substantially increased	300.21	
47. **(b)** 20a	600.5(B1)	
48. **(b)** 6"	300.14	
49. **(d)** soldered	394.56	
50. **(b)** 20a	600.5A	
51. **(c)** 6"	T. 230.51(C)	
52. **(b)** #4	314.28(A)	
53. **(b)** 12" = 1'	225.14(D)	
54. **(c)** 1" x 2"	314.23(B2)	
55. **(d)** 3 1/2"	410.38(C)	
56. **(b)** replacement	240.51(B)	
57. **(b)** metal	430.12(A)	
58. **(b)** 1 1/2 hours	700.12(A)	
59. **(c)** 15/16"	314.24	
60. **(c)** II only	220.43(B)	
61. **(d)** 10'	440.64	
62. **(a)** 5'	366.30(A)	
63. **(c)** two branch circuits	517.19(A)	
64. **(b)** #4	300.4(F)	
65. **(b)** proscenium	520.2 DEF	
66. **(b)** 16"	410.15(A)	
67. **(a)** 1 1/4"	300.4(A1)	
68. **(c)** 50v	110.27(A)	
69. **(a)** 65v	445.12(C)	
70. **(a)** 1/2"	314.24	

ANSWERS MASTER OPEN BOOK EXAM #13

1.	**(c)** 6'	410.67(C)	36.	**(b)** cleat type	410.3 ex.
2.	**(c)** 2 sq.ft.	250.52(A6)	37.	**(c)** 70%	220.61(B1)
3.	**(d)** I, II & III	430.82(B)	38.	**(c)** mechanical means	404.6(B)
4.	**(c)** 12 1/2%	250.102(C)	39.	**(d)** I, III & IV	424.19
5.	**(a)** without	312.2(A) ex.	40.	**(b)** installation	215.5
6.	**(b)** #10	Table 250.122	41.	**(d)** I, II or III	250.56
7.	**(d)** not accessible	250.68(A) ex.	42.	**(b)** 12'	Table 344.30(B2)
8.	**(b)** 30a	240.51(A)	43.	**(a)** 12"	470.3
9.	**(d)** both	440.41(B)	44.	**(d)** any of these	406.9(B4)
10.	**(c)** both I & II	210.4 ex.1,2	45.	**(c)** outline lighting	600.4(A)
11.	**(c)** III only	Table 400.4	46.	**(c)** watts	430.7
12.	**(b)** tension	400.10	47.	**(c)** 0.030a	250.162(A) ex.3
13.	**(d)** none	Table 220.12	48.	**(c)** 36"	324.10(H)
14.	**(c)** 125%	645.5(A)	49.	**(b)** II only	670.2
15.	**(c)** locked	430.102(B) ex.	50.	**(d)** receptacles	285.6
16.	**(a)** grounded	250.22(1)	51.	**(b)** less per AHJ	430.26
17.	**(c)** movement	392.8(D)	52.	**(d)** 75%	376.56
18.	**(b)** dust	500.5(C)	53.	**(d)** continuous	620.61(B2)
19.	**(b)** header	372.5	54.	**(d)** first fitting	514.9(A)
20.	**(d)** nearest entrance	230.70(A1)	55.	**(c)** #10	810.21(H)
21.	**(c)** 4'	424.59 FPN	56.	**(a)** 2'	682.2 DEF
22.	**(a)** hazardous	332.10(7)	57.	**(b)** .0293	Chapter 9 Table 5
23.	**(c)** 5 1/2'	210.52	58.	**(b)** 6'	250.102(E)
24.	**(d)** #17	Table 810.16(A)	59.	**(d)** not required	250.68 ex.
25.	**(a)** #14	650.6(A)	60.	**(b)** #18	402.6
26.	**(d)** 5/8"	501.15(C3)	61.	**(b)** galvanic action	344.14
27.	**(a)** 3.291"	Chapter 9 Table 4	62.	**(c)** I & II only	427.2 FPN
28.	**(c)** dry	380.2	63.	**(c)** GFCI	600.10(C2)
29.	**(c)** I or III	280.4(A)	64.	**(d)** electric furnace	250.21(1)
30.	**(c)** I, II & IV	680.27(A2)	65.	**(b)** 1"	366.100(E)
31.	**(d)** all	DEF 100	66.	**(a)** one more	680.24(D)
32.	**(b)** Class I, Div.II	Table 515.3	67.	**(a)** energized	300.31
33.	**(d)** I, II and III	590.6(B3a,b,d)	68.	**(a)** mogul	410.53
34.	**(c)** signal	422.42	69.	**(c)** nonlinear	220.61(C2)
35.	**(c)** III only	250.178	70.	**(c)** assembled sections	314.16(A)

ANSWERS MASTER OPEN BOOK EXAM #14

1. **(b)**	locking & grounding type	555.19(A4a)
2. **(d)**	I, II & III	680.26(A2,5) • 680.27(A1)
3. **(d)**	shall not	605.5(C)
4. **(b)**		620.91(C)
5. **(c)**	3/8"	T. 430.12(C1)
6. **(a)**	.0625"	314.40(B)
7. **(a)**	3 gallons	460.2(A)
8. **(c)**	accessible only	700.25
9. **(c)**	eff. insulated	250.112(B)
10. **(c)**	III only	392.5(E)
11. **(d)**	35%	310.15(B2a)
12. **(c)**	shielded	400.31(B)
13. **(c)**		680.25(B)
14. **(c)**	#6	800.100(D)
15. **(b)**	wooden floors	430.14(B) ex.
16. **(d)**	I & II only	600.6
17. **(c)**	is covered	640.1
18. **(c)**	6'	Table 352.30(B)
19. **(d)**	200%	630.12(B)
20. **(c)**	2v - 1.2v	480.2
21. **(d)**	I, II & III	702.2 FPN
22. **(c)**	sized to	T.250.122 •675.13
23. **(a)**	may be larger	250.122(A)
24. **(d)**	I, II & III	551.2 FPN
25. **(d)**	2000a	366.23(A)
26. **(b)**	equip. grd.	515.8(C)
27. **(b)**	terminals	200.10(A)
28. **(c)**	1"	312.11(A3)
29. **(a)**	24"	110.26(C2)
30. **(a)**	electrolytic cell	668.2 DEF
31. **(b)**	8'	300.5(D)
32. **(d)**	#20	620.12(A2)
33. **(c)**	I & III only	610.21(F4)
34. **(d)**	18"	314.23(F)
35. **(a)**	dressing rooms	520.73
36. **(c)**	functionally associated	725.26(B1)
37. **(d)**	capacitance	705.40 FPN2
38. **(b)**	II only	725.55(A)
39. **(b)**	48...60	424.72(B)
40. **(b)**	2"	362.20(B)
41. **(b)**	8"	547.5(B)
42. **(b)**	no heating	300.20(B) FPN
43. **(b)**	2	210.70(A2)
44. **(b)**	16 1/2w	424.44(A)
45. **(a)**	main bonding jumper	408.3(C)
46. **(c)**	two hours fuel	700.12(B2)
47. **(b)**	4'	422.16(B2)(2)
48. **(d)**	I, II & III	230.56
49. **(b)**	arms & stems	410.28(C)
50. **(c)**	gasoline pump	514.11(A)
51. **(d)**	6'	820.44(F3)
52. **(d)**	I, II & III	250.184(C1a,b,c)
53. **(a)**	adequate bonding	250.116 FPN
54. **(c)**	I & III only	400.22(A)
55. **(c)**	200% = twice	600.6(B)
56. **(c)**	shielded	310.6
57. **(b)**	50%	660.6(A)
58. **(d)**	all of these	250.30(4) ex.2
59. **(c)**	own section	540
60. **(d)**	I, II & III	669.9 (1)(2)(3)
61. **(a)**	8'	250.56 FPN
62. **(a)**	1000v	410.80(B)
63. **(a)**	ungrounded	685.12
64. **(a)**	bare copper	338.100
65. **(a)**	I only	410.86
66. **(a)**	surge arrester	280.2
67. **(c)**	overload	430.55
68. **(b)**	163a	Table 326.80
69. **(b)**	125...250	440.55B
70. **(c)**	both	390.6

ANSWERS MASTER OPEN BOOK EXAM #15

1. **(b)** Class I Div. II 511.3(B1)
2. **(a)** 6' 250.56
3. **(b)** 75a 430.110(A) • T.430.250
4. **(d)** not at all 314.20
5. **(d)** none of these 650.8
6. **(d)** I,II,III & IV 450.11
7. **(c)** 100% 210.21(B1)
8. **(c)** intrinsically safe 504.80(C)
9. **(b)** discharging 280.2 DEF
10. **(d)** all of these 620.11(A)
11. **(c)** 6' 250.118(6c)
12. **(d)** 115% 440.12(A1)
13. **(c)** neutral 250.166(A)
14. **(a)** Class I Div.I 516.2(B5)
15. **(a)** 90 days 590.3(B)
16. **(d)** removed 390.7
17. **(a)** 5 feet 680.71
18. **(d)** all 410.24
19. **(d)** I, II and III 600.9(D1,2,3)
20. **(a)** ozone 310.6
21. **(d)** any 250.119
22. **(a)** 300v 386.12(2)
23. **(d)** I, II & III 725.82(L) FPN
24. **(c)** Class III Div. II 500.5(D2)
25. **(b)** simul. 404.2(B) ex.
26. **(c)** 12' 210.62
27. **(c)** 50° 424.36
28. **(c)** 800a 240.4(B3)
29. **(d)** I, II and III 210.11(C1,2,3)
30. **(d)** second ground fault 517.19(F) FPN
31. **(c)** 90% T. 555.12
32. **(d)** Stranded 410.28(E)
33. **(b)** 660w 720.5
34. **(c)** 10v 517.64(A1)
35. **(c)** lighting track 410.100
36. **(b)** corrosion prot. 344.10(B)
37. **(a)** 400 sq. ft. 552.2 DEF
38. **(b)** 8" 382.30
39. **(c)** safeguarding 90.1(A)
40. **(d)** I,II & III•398.30(A1)•398.42•404.10(A)
41. **(d)** all of these 770.6
42. **(b)** service entrance 250.102(C)
43. **(a)** 12" 450.22
44. **(a)** 1 Table 348.22
45. **(d)** I,II & III 340.116
46. **(d)** liquid-tight 502.10(A2)(2)
47. **(c)** time delay 440.54(B)
48. **(d)** not allowed 502.100(A3)
49. **(a)** isolating 404.13(A)
50. **(d)** GFCI 680.32
51. **(b)** 3 1/2' Table 110.26(A)
52. **(d)** .8 Chapter 9 Table 1 Note 7
53. **(c)** not exceed 600v 430.111(B3)
54. **(b)** #6 280.23
55. **(c)** 2" 600.9(C)
56. **(b)** protector 800.90(B) FPN
57. **(b)** II only 240.8
58. **(a)** 1/4" 600.41(C)
59. **(b)** 115% 445.13
60. **(b)** fixed 400.8 (1)
61. **(c)** 3/4" 250.52(A5a)
62. **(d)** 125% 520.25(A)
63. **(b)** MI 518.4
64. **(d)** all 368.238
65. **(b)** 1/4" 300.6(C)
66. **(c)** 100a 430.72B ex.1 • T.430.72(B) Col.B
67. **(b)** hoistways 392.4
68. **(b)** 24" = 2' 514.8 ex.2
69. **(c)** #8 310.3
70. **(d)** 100a 550.32(C)

ANSWERS MASTER OPEN BOOK EXAM #16

1.	**(d)** 1/4"	312.2(A)
2.	**(d)** not required	314.23(E)
3.	**(c)** within sight •600.6A•DEF 100 within 50'	
4.	**(b)** 72°F normal	402.3 FPN
5.	**(d)** I,II, III	390.2(A)
6.	**(b)** release-type	324.10(H)
7.	**(b)** I and II	300.22(C2)
8.	**(c)** racked	110.12(B)
9.	**(a)** 65a	310.15(B2a) ex.4
10.	**(c)** 7	430.9(C)
11.	**(c)** 2 x 4	410.15(B1)
12.	**(a)** both	350.10
13.	**(d)** four	517.18(B)
14.	**(c)** both ends	280.25
15.	**(b)** frame DEF .053"	551.2 DEF
16.	**(d)** 15watts sq.ft.	424.99(B)
17.	**(b)** accidental	430.73
18.	**(d)** I or II	430.109(A1,2)
19.	**(b)** 0.06	250.52(A6)
20.	**(d)** either	460.28(B)
21.	**(c)** I & II only	700.16
22.	**(d)** neither	250.148(B)
23.	**(d)** neither	501.130(A3)
24.	**(c)** panelboard	408.36(A) ex.1
25.	**(b)** cable trays	336.10(2)
26.	**(d)** one minute = 60 sec	701.11
27.	**(d)** I,II,III & IV	300.11(1,2) ex.
28.	**(c)** 6 times	240.101(A)
29.	**(c)** 24"	T.300.50 • 300.50(A) ex.2
30.	**(b)** II only	517.72(C)
31.	**(b)** II only	250.110 ex.2
32.	**(c)** 30'	366.12(2)
33.	**(b)** current-carrying	520.53(O2)
34.	**(a)** potential diff.	250.60 FPN2
35.	**(b)** vented power fuse	DEF 100 FPN
36.	**(a)** 600	362.12(6)
37.	**(d)** 2'	550.32(F)
38.	**(b)** below	210.52 FPN
39.	**(d)** I, II, III	511.12
40.	**(b)** readily accessible	450.13
41.	**(b)** staggered	520.65
42.	**(d)** 4 1/2'	334.30
43.	**(a)** b.c. rating	210.20(A)
44.	**(b)** simultaneously	410.48
45.	**(d)** short as possible	517.19(C) FPN
46.	**(b)** frames insulated	250.110 ex.1
47.	**(c)** 4"	390.3(A)
48.	**(d)** yellow stripe	200.6(D)
49.	**(b)** I & III only	513.7(B)
50.	**(c)** Article 504	505.15(A)
51.	**(c)** both I & II	250.52(A1) (D1)
52.	**(c)** 7"	Table 490.24
53.	**(b)** 4.308 sq. in.	Table 384.22
54.	**(a)** 18"	760.52(B1)
55.	**(d)** #12	720.4
56.	**(b)** 194° F	320.80(A)
57.	**(d)** barrier	810.18(C)
58.	**(b)** interchangeable	410.105
59.	**(b)** same grd. electrode	250.58
60.	**(b)** 300v	404.8(B)
61.	**(d)** 3	Table 430.37
62.	**(a)** #14	310.14
63.	**(c)** 12'	511.7(B1b)
64.	**(d)** cable assembly	230.30 ex.4
65.	**(c)** both I & II	430.225(C1a)
66.	**(d)** 200%	630.12(A)
67.	**(d)** neither I nor II	501.30A
68.	**(c)** both I & II	250.24(A1)
69.	**(a)** 31%	Chapter 9 Table 1
70.	**(d)** I,II or III	250.94(1,2,3)